Shallow Refraction Seismics

Series Editor

D. S. Parasnis

Professor of Applied Geophysics, University of Luleå, Sweden
Fellow of the Royal Swedish Academy of Engineering Sciences

Shallow Refraction Seismics

Bengt Sjögren

LONDON   NEW YORK
Chapman and Hall

First published 1984 by
Chapman and Hall Ltd
11 New Fetter Lane, London EC4P 4EE
Published in the USA by
Chapman and Hall
733 Third Avenue, New York NY10017

© 1984 Bengt Sjögren

Printed in Great Britain at the
University Press, Cambridge

ISBN 0 412 24210 9

British Library Cataloguing in Publication Data

Sjögren, Bengt
  Shallow refraction seismics.
  1. Seismic waves
  I. Title
  551.2′2   QE538.5

  ISBN 0-412-24210-9

Library of Congress Cataloging in Publication Data

Sjögren, Bengt
  Shallow refraction seismics.

  Bibliography: p.
  Includes index.
  1. Seismic refraction method.   I. Title.
  TN269.S536   1984   622′.159   83-25168
  ISBN 0-412-24210-9

Contents

Preface

There are many general geophysical textbooks dealing with the subject of seismic refraction. As a rule, they treat the principles and broad aspects of the method comprehensively but problems associated with engineering seismics at shallow depths are treated to a lesser extent. The intention of this book is to emphasize some practical and theoretical aspects of detailed refraction surveys for civil engineering projects and water prospecting.

The book is intended for students of geophysics, professional geophysicists and geologists as well as for personnel who, without being directly involved in seismic work, are planning surveys and evaluating and using seismic results. The latter category will probably find Chapters 1, 5 and 6 of most interest.

Interpretation methods, field work and interpretation of field examples constitute the main part of the book. When writing I have tried to concentrate on topics not usually described in the literature. In fact, some discussions on interpretation and correction techniques and on sources of error have not been published previously. The field examples, which are taken from sites with various geological conditions, range from simple to rather complicated interpretation problems.

Thanks are due to A/S Geoteam (Norway), Atlas Copco ABEM AB (Sweden), BEHACO (Sweden) and the Norwegian Geotechnical Institute for allowing me to use field examples and certain data from their investigations.

I should particularly like to thank Professor Dattatray S. Parasnis of the University of Luleå (Sweden) for revising the manuscript and for his numerous invaluable suggestions.

Tvåspannsv. 20   Bengt Sjögren
Järfälla, Sweden

1

Introduction

The history of the seismic refraction method goes back to 1910, when the German geophysicist L. Mintrop pointed out the practical use of the transmission of seismic waves through the earth. In 1919 Mintrop applied for a patent covering refraction profiling to determine depths and types of subsurface formation. At the same time the principles of the method were analysed in the USA and put into practical use by E. V. McCollum, to mention but one worker on the subject. By 1925 the refraction method was well established as a tool in applied geophysics. In the early days the method was used for oil exploration and for detecting hidden salt domes. At the beginning of the 'thirties the refraction technique was also seen to be applicable to civil engineering problems.

In 1941 the first seismic depth-to-bedrock investigation was carried out in Sweden for a planned hydroelectric power plant, thereby initiating extensive use of the method in Scandinavia, mainly for civil engineering and to a lesser extent for water prospecting and mining. Progress was rapid in refinement of instrumentation as well as in field and interpretation techniques. The method proved to be reliable and inexpensive, and from the start the demand for this type of investigation increased rapidly. From being the exception, before long the application of the method became routine in connection with civil engineering problems, mainly related to power schemes. The widespread use of seismics in Scandinavia at a very early stage is due to a fortunate coincidence of events and conditions. The introduction of the method, using reliable and high-capacity instruments developed by the ABEM Company,* came just at the time when hydroelectric power systems were being built up on a large scale. As the constructions went underground, prior information on the subsurface became vital.

The applicability of the refraction method increased considerably from the

*The ABEM Company/Terratest AB ceased to exist in 1977 as a contracting company for geophysical investigations. Manufacturing of geophysical instruments has been continued by the present Atlas Copco ABEM AB, Bromma, Sweden.

beginning of the 'fifties owing to refinements in the interpretation technique. The accuracy in the depth determinations was improved by the introduction of special correction methods which took into account the interpretation problems connected with greatly varying geological and topographical conditions. The measurements had to be restricted earlier to rather flat ground, but from then onwards they could also be carried out in very irregular terrain. The next advance came in 1956 when a new interpretation technique for velocity determinations enabled detailed predictions of rock quality to be made. The advantages of the developments in interpretation could be utilized fully owing to the high-resolution instruments that came into use around 1951.

The seismic method utilizes the propagation of elastic waves (sound waves) through the earth and is based on the following fundamental postulates:

1. The waves are propagated with different velocity in different geological strata.
2. The contrast between the velocities is large and
3. The strata velocities increase with depth.

Departures from the postulated velocity conditions will make it difficult or even impossible to use the method. Fortunately, however, the departures occur rarely.

The energy needed to generate the seismic waves is obtained by detonating small explosive charges or by weight dropping. The arrival of the waves at the receiving stations after propagation through the earth is recorded by detectors sensitive to vibration. If the distances and travel times between the impact point and the receiving stations are known the velocity of a wave in a particular layer can be determined.

The seismic field work is generally carried out with the impact points and detectors placed in a straight line, which is the so-called in-line profiling system. It is of advantage to the field work as well as to the interpretation to keep a regular distance between the detector stations. For measurements on land the detectors used (called geophones or seismometers) are sensitive to ground vibrations, while those used under water (hydrophones) are sensitive to variations in water pressure. The impact of the ground motion (or pressure variations) on the detectors is transformed into an electric current and the signal (current) is transmitted by cables to a seismograph which comprises an amplifying unit and a recorder. The signals are recorded either on photographic film or on magnetic tape. The instant of the explosion or of the impact from a mechanical energy source is conveyed by a cable or via radio to the recording equipment. In shallow refraction surveys the distances between the receiving stations are kept small, generally 5 or 10 m. The term shallow refers to the type of project and not to the refraction method as such. The depth of interest in a civil engineering project seldom exceeds 100 m.

The refraction method makes use of waves travelling along the ground surface and of waves in the underlying more compact layers where the velocities are higher. The waves in the subsurface strata return to the ground surface as refracted waves which are sometimes also called head waves. In the vicinity of the impact point the ground surface waves are the first to arrive. At a certain distance from the impact point, the waves following the longer but faster paths in the underlying layers overtake the surface waves. The distances from the impact point to the points where the refracted waves are recorded as first arrivals are a function of the strata velocities and depths.

The interpretation ultimately giving the depths and velocities can be carried out manually or by data processing. The reliability in the determination of the depths and layer velocities increases if the signals are recorded in both directions along the detector layout (spread).

A seismic survey yields considerable information for a variety of projects. Usually only the longitudinal velocities based on the first arrivals of compressional waves are used for the interpretations, but in the evaluation of the subsurface conditions the tendency at present is also to include other parameters such as transverse velocities (shear waves), elastic constants, amplitudes and frequencies.

The following is a summary of the type of information generally obtained by a conventional seismic investigation, i.e. when only the first arrivals of the longitudinal waves have been used for the interpretation.

1. Thicknesses of the overburden layers overlying compact bedrock. The velocities in these layers give some indication of the material composition as well as the degree of packing and water content. The level of a groundwater table can be revealed by a sudden velocity increase. Sometimes intermediate layers, harder or softer than the surrounding layers, are encountered. Such layers may be apparent from velocity anomalies. In serious cases inverse velocity relations may invalidate the depth interpretations. But, as mentioned previously, in a layer sequence the velocities generally increase with depth.

   The actual geology must be taken into account in correlating velocity layers and geological formations. In one area, a certain velocity may correspond to a hard, water-saturated soil layer, while the same velocity may correspond in another case to a fissured, weathered upper rock layer.

2. The total depth down to the bedrock, and, in the case of surficial layers of fractured and/or weathered rock material, the depth of compact rock. Since the depths are calculated at the impact points as well as at the receiving points, a continuous and detailed bedrock relief is obtained along the entire traverse investigated. The continuous picture of the subsurface is a distinctive feature of the seismic method and this condition reduces the risk that sections that are critical or interesting for a project remain undetected.

3. The quality of the rock as defined by the velocity in it. For most purposes the interest is focused on sections where lower velocities indicate rock material of inferior quality. The lower velocities appear as zones along the

bedrock surface or as horizontal beddings. The former velocities indicate the presence of faults, fractured zones, contact zones, deep depressions in the bedrock surface, another looser rock type, etc., while in the latter case the velocities correspond to upper layers of weathered and fractured or softer rock material. Note that a soft rock material, overlain by a harder rock layer with a higher velocity, cannot generally be detected by refraction seismics.

The actual geology must be taken into account in using the velocities as a measure of rock quality: a low velocity obtained in highly fractured granite may in another area correspond to a comparatively compact, young sandstone.

In the early days of the application of refraction seismics in Scandinavia the surveys were mostly carried out for power projects but the scope has gradually widened with the constructing engineers' increased knowledge of, and confidence in, the method. A list of situations in which refraction seismics is vital will be rather long but some occurring relatively frequently are as follows:

1. Underground constructions to be placed in rock. Among these can be mentioned machine halls for power stations, tunnels and their entrances, oil and petrol storage depots, air raid shelters, military installations, factories, mines and sewage treatment plants. Seismic surveys are used in such cases to determine the depth of solid bedrock to obtain an estimate of the rock cover available for the construction. Just as vital is to get information on the rock quality as indicated by the seismic velocities. The principal aim of the investigations is to find compact rock not affected by systems of shear zones. Rock sections with low seismic velocities often constitute the most critical parts in a project area, causing great problems when the project is being realized, especially if the excavation work has to be carried out under water pressure. If weak rock sections cannot be avoided, for example when a tunnel has to intersect a shear zone, seismics is used to find the best alternative for the tunnel. Besides, since the rock conditions are known in advance, the planning of the tunnel driving is facilitated and the risk of encountering unforeseen problems is reduced.

2. Constructions to be founded on rock or on other solid layer. In this category we may include bridges, dam sites, heavy industrial buildings, nuclear power plants, harbour quays, etc. It is of importance here to obtain the depth to the bedrock or other solid layer, on which the construction can be founded. The seismic rock velocities are used for evaluating the risk for possible future water leakages under dam constructions or pollution of the groundwater. Water flow may occur in the more or less vertical shear zones in the bedrock, in upper, weathered and/or fissured rock layers or in intermediate, permeable soil layers. The term soil is used here for the accumulation of unconsolidated mineral particles, partly mixed with

organic matter, which covers the bedrock. Karst phenomena, sink holes, in limestone create special problems for dam constructions, since the cavities are often very limited in size. Even if the seismic profiles are spaced closely, the sink holes may remain unrevealed.

3. Another group of projects is composed of those where rock or other harder layer is undesirable, for instance, harbour basins and accesses, canals, channels, roads, railways. Of interest for these projects is the level of more solid layers, bedrock or harder overburden. The velocities can be used to estimate the rippability and excavatability of rock and soil material. Harder intermediate layers, indicated by velocities higher than those in the over- and underlying soil material, have proved problem-causing when dredging.

4. Landslides and erosion problems. During the last decade the refraction method has been increasingly applied to this kind of problem, often in combination with geotechnical methods. Thicknesses and velocities of the soil layers and the overall picture of the subsurface aid the evaluation of the risks for landslides or soil erosion. Geotechnical data, such as the strength of a material and its composition, can be extended over a greater area with the help of the associated seismic velocities.

The refraction method has also had a wide application in the search for resources, often in connection with civil engineering projects. Examples are:

1. Selection of sand and gravel deposits. By an adequate combination of seismics and sampling an estimate can be obtained of the available masses containing material suitable for the desired purpose.

2. Quarry sites. The seismically determined depths-to-bedrock are used to evaluate the exploitation feasibilities and costs.

3. Water prospecting. Groundwater can be looked for in the overburden or in the bedrock.

   A sudden velocity increase in a soil layer can indicate a groundwater table (or a harder layer). In doubtful cases test drillings should be made. If transverse velocities have been obtained, the problem can be solved by seismics since transverse waves – unlike longitudinal waves – are not affected by varying water content. The same transverse velocity recorded above or below an interface – indicated by an increase in the longitudinal velocity – shows that the interface corresponds to a water table, while an increase in the transverse velocity below the interface also proves the existence of a harder layer.

   The volume of the water-bearing strata in the overburden can be calculated using the seismically determined depths. The velocities give some indication of the permeability of the various layers. If the seismic measurements are sufficiently comprehensive, contour lines of the groundwater table may be established and the direction of the water flow can be determined.

In bedrock the low-velocity zones give indications of fissured, water-bearing sections. A seismic survey cannot prove the occurrence of water as such but is indispensable for tectonic analysis, particularly in areas where the bedrock is covered by relatively thick layers of overburden. However, the flow-rate has proved 5–10 times larger in water wells located on seismic indications than in those placed at random.

Sedimentary rock formations and fractured and weathered upper rock layers with a more or less horizontal bedding can be mapped by seismics. Whether they are water bearing or not is difficult to say. The velocity pattern is often complex since the magnitude of the velocities depends on the degree of saturation and the porosity of the rock. Sedimentary strata, known to have a high content of water, can be traced by seismics.

4. Alluvial mineral deposits. The volume of layers containing minerals can be obtained by seismics. Exploratory drilling can be directed towards the depressions in the layers, buried channels, etc., where there is often an accumulation of minerals. The excavation and dredging operations are facilitated owing to the continuous picture of the subsurface given by the seismic investigation.

5. Zones of weakness in the bedrock, where mineralization occurs in fissures and fractures or zones where the weathering has been intensified in rock sections with a higher mineral content, so-called gossan plugs. The extent and direction of such zones can be traced by a refraction survey.

The aim of a seismic investigation is to obtain data of a technical nature, but the economic aspects of the method are also of great importance. Some of these are:

1. The method is rapid and alternative solutions to a project can be evaluated in a very short time so that the time for the planning of a project is reduced.

2. An immense amount of data covering large areas of a project site is obtained at a reasonable cost. A seismic investigation yields the depths at the impact points as well as at the detector stations. If we assume a depth to the bedrock of 10 m, and the detector and impact separations are 5 and 25 m respectively, the total calculated depth is 250 m per 100 m of profile length. In the case of top layers of fissured and/or weathered rocks, these will be included in the depth results. Moreover, the rock quality for the entire profile length measured will have been indicated by the waves travelling along the bedrock. The economic advantage increases with increasing depth of investigation, since the measuring rate is almost independent of the depth. For an average depth of 50 m, the total sounding depth is 650 m per 100 m of profile length. The detector and impact distances are in this case assumed to be 10 and 50 m respectively.

3. The total investigation costs are reduced, since more time-consuming and expensive investigation methods such as drilling can be directed, according to the seismic results, towards the most critical or interesting parts of a

project, thereby also increasing the value of the information obtained by the drilling operations. It is advantageous to adapt the drilling to the seismic results and not to follow a drilling programme fixed rigidly in advance.

4. Tendering for a project is facilitated owing to the more complete information obtained and claims due to unforeseen geological conditions can be avoided to a large extent.

One vital question is the expected accuracy of the seismic results. When accuracy is discussed, the term generally refers to the depth determinations. However, the reliability of the seismic velocities should also be included in an accuracy analysis, since they are used for evaluation of rock quality and material composition. This is discussed further in Chapter 2.

In general, it can be stated that the seismic results are adequate for a great variety of projects, but allowances have to be made. The accuracy of seismic results is not an unequivocal concept. It depends, for instance, on the actual geology, the position of the seismic lines in relation to dips of layers and the rock structure, the field and interpretation procedures employed, and knowledge and experience of the personnel involved in the work. The reliability of the seismic results usually increases with an increasing amount of measuring material since more information is available for the evaluation of velocities and depths. However, it should be noted that the accuracy has no value in itself; it must always be related to the aim of the investigation and the type of project. The reliability demand is not the same for a reconnaissance survey looking for sand deposits over a large area as for a detailed investigation of a critical part of a tunnel project.

Statistical comparisons between depth-to-bedrock determinations obtained by drilling and by seismics have shown a mean difference of about ± 1 m for an overburden depth of 10 m. For greater depths the divergences have proved less than $\pm 10\%$ of the drilling depth. These figures cannot, however, be regarded as the accuracy of the seismic results, since other sources of error than those inherent in the seismic method are included in the comparisons. The divergences include errors in the seismic work as well as in the drilling operations and in the mapping of the drill holes and the seismic profiles. Discrepancies between depths obtained by seismics and by drilling are, however, in general to be expected because a drill hole samples a single spot, while seismics gives an average depth of the immediate vicinity below the impact point or detector station.

It is obvious that seismic results are checked by drilling but the reverse is less obvious. A drill hole gives detailed and accurate information but for a very limited area and volume, so that depths and rock quality in the vicinity of the drill hole remain completely unknown parameters. Results from drill holes placed at random may not be representative of the general geological conditions of a project site. This can be revealed by comparing the drilling results with the overall picture – depths and rock quality – obtained by the seismic survey.

If the aforementioned conditions concerning the fundamental principles for the method, such as velocity increase with depth, etc. are not fulfilled, errors in depth can be grave. Sources of error are to be found in:

1. Absence of velocity contrast between different types of material. (A water-saturated soil layer can have the same velocity as an underlying extremely weathered and fractured sedimentary rock layer.)
2. Inverse velocity relations. (Intermediate layers have velocities higher or lower than the surrounding layers.)
3. A hidden layer. (An intermediate layer is masked by overlying layers, i.e. the layer is not represented by first arrivals on the records.)

2

Basic principles

The propagation speed of seismic waves through the earth depends on the elastic properties and density of the materials. A stress applied to the surface of a body tends to change the size and shape of the body. The external stress gives rise to opposing forces within the body due to the deformations, the strains, of the body. The ability to resist deformation and the tendency of the body to restore itself to the original size and shape define the elasticity of a particular material.

2.1 STRESS, STRAIN AND ELASTIC CONSTANTS

The stress S is defined as the force F per unit area A, thus $S = F/A$. A stress acting perpendicularly to the area is called compressive or tensile, depending on whether it is directed into or from the body. A compressive stress tends to cause a shortening of the body, and a tensile stress an elongation. At right angles to the direction of the stress, the body dilates or contracts, depending on whether the stress is compressional or tensile. The stresses preserve the shape of the body but change the volume. The longitudinal strain ϵ_1 is defined as the ratio of the elongation or shortening Δl to the original length l of the body, thus $\epsilon_1 = \Delta l/l$. A transverse strain ϵ_w is defined as the ratio of the expansion or contraction Δw, perpendicular to the direction of the stress, to the original width w of the body, thus $\epsilon_w = \Delta w/w$.

According to Hooke's law a strain is directly proportional to the stress producing it. The statement holds for small strains only, but not if the stress is so large that it exceeds the elastic limit of a substance. However, this limitation of the law can be left out of account here.

The relation between longitudinal strain and stress is

$$\Delta l/l = \frac{F/A}{E}$$

or

$$E = \frac{F/A}{\Delta l/l} \tag{2.1}$$

which is the stress per unit area divided by the relative elongation or shortening. The constant E is called Young's modulus. Note that for our purposes E is the dynamic modulus of elasticity and not the static modulus.

A force acting parallel to a surface is called a shearing stress. Because of this the right angle between two perpendicular directions is changed to an angle $90° - \varphi$. The deformation angle φ is the shearing strain. If φ is small, it is proportional to the stress. The relationship is

$$\mu = \frac{F/A}{\varphi} \tag{2.2}$$

The proportionality constant μ, the rigidity or shear modulus, is a measure of the material's resistance to shearing strain.

Another elastic constant of interest is the bulk modulus. A uniform compressive stress acting on a body causes a decrease in the volume of the body. The stress divided by the relative volume change defines the bulk modulus k.

Thus

$$k = \frac{F/A}{\Delta V/V} \tag{2.3}$$

where V = volume and ΔV = the change in volume. The ratio $\Delta V/V$ is referred to as the dilation. The reciprocal of k is called the compressibility.

The relation between the transverse strain and the longitudinal strain, whether the stress is compressive or tensile, is called Poisson's ratio, σ. The ratio is given by

$$\sigma = \frac{\Delta w/w}{\Delta l/l} \tag{2.4}$$

Poisson's ratio lies around 0.25 for most solids.

Lamé's constants λ and μ are valid for isotropic media, i.e. media in which the elastic properties are independent of direction. Expressed in E and σ they are

$$\lambda = \frac{\sigma E}{(1+\sigma)(1-2\sigma)} \tag{2.5}$$

and

$$\mu = \frac{E}{2(1+\sigma)} \tag{2.6}$$

The Lamé constant μ is the same as the aforementioned rigidity modulus.

When Poisson's ratio is 0.25, the constants λ and μ are equal and equal to $2E/5$.

The various constants above are related to each other. Some useful relations are given below.

$$E = \frac{\mu(3\lambda + 2\mu)}{\lambda + \mu} = \frac{9k\mu}{3k + \mu} \tag{2.7}$$

$$\sigma = \frac{\lambda}{2(\lambda + \mu)} = \frac{3k - 2\mu}{6k + 2\mu} \tag{2.8}$$

Since liquids have no resistance to shear, $\mu = 0$ and $\sigma = 0.5$, which is the maximum value for σ. The values for σ vary from 0.05 to 0.45. The lower figures of σ refer to very hard material, while the higher ones are obtained in very soft, unconsolidated material.

Also

$$k = \frac{E}{3(1 - 2\sigma)} = \frac{3\lambda + 2\mu}{3} \tag{2.9}$$

When σ is 0.25, k is equal to $2E/3$. With increasing σ the value of k increases, and when k and E are equal, $\sigma = 1/3$. For σ values higher than $1/3$, k is greater than E. Since for a liquid $\mu = 0$ and $\sigma = 0.5$, $k = \lambda = \infty$.

2.2 ELASTIC WAVES

When stresses and strains are not in equilibrium, for instance when the stress applied to a medium is removed, the strain condition in the medium starts to propagate as an elastic wave. Inside a homogeneous body there are two types of wave that can be generated. The first type of body wave is variously called longitudinal, compressional or P-wave. The letter P stands for primary since these waves are the first to be detected on an earthquake record. The particle motion for these waves is to and fro in the direction of the propagation. They correspond to ordinary sound waves. The second wave type, where the motion of the particles is at right angles to the propagation direction, is called transverse, shear or S-wave. They are usually recorded as the second earthquake event. At the contact surface between two media there are also other wave types besides the above body waves. These waves, for example Rayleigh waves and Love waves, travel close to the contacts between the media. For the Rayleigh waves the particle motion is in a vertical plane and elliptical and retrograde with respect to the propagation direction. The Love waves may exist when there is a low-velocity layer underlain by a medium with a higher velocity. The wave motion is perpendicular to the propagation direction of the waves and in a plane parallel to the contact surface.

The velocities of the longitudinal and transverse waves, designated V_P and V_S respectively, are expressed in terms of the elastic constants as follows

$$V_P = \left(\frac{\lambda + 2\mu}{\rho} \right)^{1/2} = \left[\frac{k + (4/3)\mu}{\rho} \right]^{1/2} = \left[\frac{E}{\rho} \frac{1 - \sigma}{(1 + \sigma)(1 - 2\sigma)} \right]^{1/2} \qquad (2.10)$$

and

$$V_S = \left(\frac{\mu}{\rho} \right)^{1/2} = \left[\frac{E}{\rho} \frac{1}{2(1 + \sigma)} \right]^{1/2} \qquad (2.11)$$

ρ being the density.

The equations show that the velocity of the transverse waves is lower than that of the longitudinal waves in the same medium. It can also be seen that for a fluid the velocity V_S is equal to zero since the rigidity modulus μ is zero. Transverse waves cannot propagate through fluids. A consequence of this is that the transverse velocities are less affected than the longitudinal velocities by a varying degree of moisture content.

The relation between the velocities is

$$\frac{V_S}{V_P} = \left(\frac{1/2 - \sigma}{1 - \sigma} \right)^{1/2} \qquad (2.12)$$

When σ increases from, say, zero to its maximum value 0.5, the ratio V_S/V_P decreases from $1/\sqrt{2}$ to zero. The equation also shows that transverse velocities vary from zero (in a liquid) to a maximum of 0.7 of the longitudinal velocity for a solid for which $\sigma = 0$. For an assumed σ value of 0.25 for hard rocks, the ratio is $1/\sqrt{3}$, which means that V_S is 58% of V_P.

Figure 2.1 shows some empirical relations between longitudinal and transverse velocities. The average line in the figure is based on measurements on 93 rock sections from five sites with igneous and metamorphic rocks. The longitudinal velocities vary in this case from 3300 to 5700 m/s and the transverse from 1600 to 3400 m/s in the various rock sections. Note that the velocities are shown in kilometres per second in the figure. Roughly speaking, the transverse velocities vary on average from about 53 to 55.5% of the longitudinal velocities. The former figure refers to the lower and the latter figure to the higher velocity range. The dispersion, expressed in V_S, of the individual values around the average curve in the figure is 175 m/s. The scatter of the values is related to the variations in Poisson's ratio σ. With increasing or decreasing σ, the values diverge from the average. Rocks having low σ lie above the average and those with high σ lie below. At one site the departure of V_S from the average was only 65 m/s because of a small dispersion of the σ values. The results from drilling indicated rather homogeneous rock conditions. On the other hand, comparisons at other sites showed that an

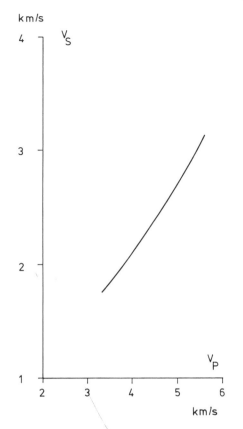

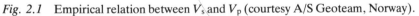

Fig. 2.1 Empirical relation between V_s and V_p (courtesy A/S Geoteam, Norway).

increased dispersion of the seismic constants in relation to the average values corresponds to more irregular geological conditions with high fracture frequencies.

The relative occurrence of longitudinal (1) and transverse (2) velocities is illustrated in Fig. 2.2 by means of statistical distribution and cumulative curves. The curves are based on 4100 m of detailed in-situ velocity determinations of igneous and metamorphic rocks.

The velocity of the Rayleigh waves is lower than that of the body waves in the same substance, being about nine-tenths of the transverse velocity. The Love waves are essentially shear waves. According to Love the waves propagate by multiple reflections between the top and the bottom surface of the layer, and for short wavelengths their velocity is equal to the transverse velocity in the upper layer and for long wavelengths to the transverse velocity in the underlying layer. The surface waves have found little application in applied refraction work, mainly because they do not penetrate deep.

% of total profile length

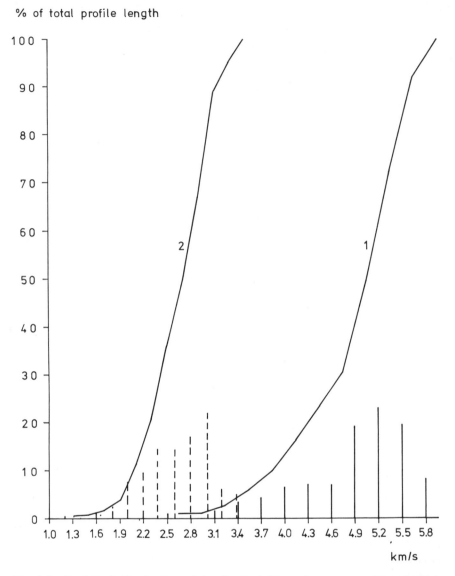

Fig. 2.2 Statistical distribution of longitudinal (1) and transverse (2) velocities (courtesy A/S Geoteam, Norway).

2.3   WAVELENGTH AND FREQUENCY

A compressional wave travelling through a medium gives rise to alternating condensations and dilatations. The distance between two wave phases of the same kind is the wavelength, usually represented by the symbol λ. The period

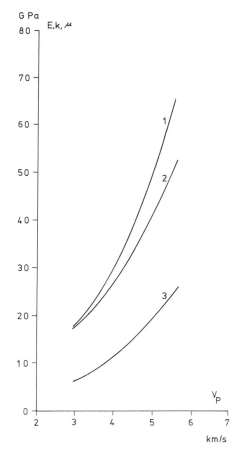

Fig. 2.3   E, k, μ as functions of V_p (after Sjögren *et al.* 1979).

T is the time required for a wave to travel the distance of one wavelength. The frequency f is the reciprocal of the period, thus $f = 1/T$, measured in cycles per second (c/s) or in hertz (Hz). Since $\lambda = VT$, the wavelength, velocity and frequency are related by the equation $\lambda = V/f$. The frequency of the surface waves is low, generally from 10 to 15 Hz. The body waves cover a band from about 15 to about 100 Hz.

The moduli E, k and μ are plotted in Fig. 2.3 as functions of the longitudinal velocities, (1): E, (2): k and (3): μ. The average curves are based on 80 examples from three igneous and metamorphic rock areas. The calculated values for the moduli diverge from the average curves with increasing or decreasing Poisson's ratio σ. As regards E and μ, values with high σ are to be found above, and those with low σ below, the average. The reverse is true for k.

2.4   LONGITUDINAL VELOCITIES

Almost exclusively the longitudinal velocities have been used in refraction work for depth determination and for evaluation of material composition and rock quality. One reason for the dominant use of the longitudinal waves in seismic exploration is that they are the first to reach the detector stations and so tend to obscure later arrivals. However, there is now an increasing tendency to exploit the entire potential of the measurements by including evaluations based on transverse waves, elastic constants, amplitude and frequency analysis. The new equipment with tape recording and filtering possibilities will probably increase the applicability of the method.

Some typical longitudinal velocities measured in situ are presented in Fig. 2.4. The value of the velocity for a given material covers a considerable range. The shadowed parts indicate the more frequently occurring ranges. There are some characteristic features of the velocity distribution:

(a) The great velocity contrast between the unconsolidated soil materials and the igneous/metamorphic rocks.
(b) The transitional nature of the velocities in the younger sedimentary rocks.
(c) The influence of the water content on the overburden velocities.
(d) The partial overlapping of velocities for different types of material.

Velocities lower than that of sound waves in air can be encountered in dry and loose soil layers. The reason is that the higher density of the soil material is not compensated for by the increase in the elastic constants. Water saturated soil layers may for the same reason display velocities lower than the velocity in water, which is generally 1450–1500 m/s. Layers with a high content of organic matter have sometimes given velocities of 600–800 m/s when below a groundwater table.

It is evident from Fig. 2.4 that the velocities in sedimentary rock formations increase with age. The reason is that the older sediments have been subjected to metamorphism. The velocities of sediments also increase with depth of burial while the porosity decreases because of the increased pressure.

The velocity values in Fig. 2.4 refer to relatively competent rocks. In the case of an increased weathering and/or fracturing of the rock material, considerably lower velocities can be obtained. In Fig. 2.5 the longitudinal velocities have been correlated with the fracturing frequency, expressed in Rock Quality Designation (RQD) values and in the number of cracks per metre. The fracturing data were obtained in drill holes located close to or on the seismic profiles. Since in shallow refraction surveys the wave penetration is limited, only samples from the upper 25 m of the bedrock have been included in the comparisons. Curve 1 shows the correlation between velocities and the number of cracks per metre while curve 2 refers to the fracturing expressed in RQD values for the same rocks. The curves are based on drilling and seismic data from 74 rock sections. The correlation curve 3

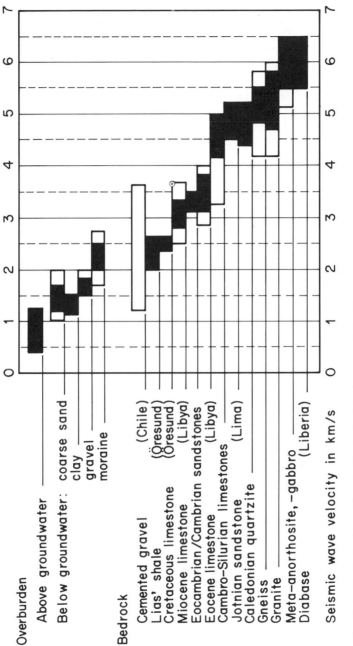

Overburden

Above groundwater

Below groundwater: coarse sand
clay
gravel
moraine

Bedrock

Cemented gravel (Chile)
Lias' shale (Öresund)
Cretaceous limestone (Öresund)
Miocene limestone (Libya)
Eocambrian/Cambrian sandstones
Eocene limestone (Libya)
Cambro–Silurian limestones
Jotnian sandstone (Lima)
Caledonian quartzite
Gneiss
Granite
Meta–anorthosite, –gabbro
Diabase (Liberia)

Seismic wave velocity in km/s

Fig. 2.4 Typical values of longitudinal velocities (courtesy Atlas Copco ABEM/AB (Sweden)).

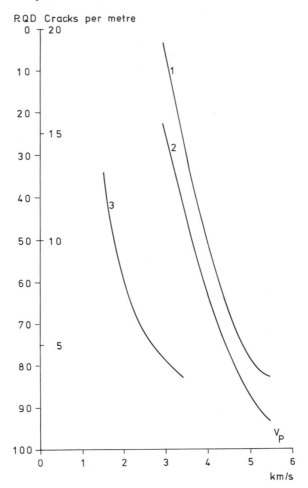

Fig. 2.5   Fracturing as a function of V_p.

shows RQD values for Permian and Triassic sandstones. The deviations of the individual examples from the average curves are rather moderate. They range from about 1.0 crack per metre for the higher velocities to 1.5–2.0 cracks per metre for the lower velocities. The corresponding dispersions of RQD values are 2–3% and 5–6%.

In anisotropic media the recorded velocities are generally higher when measured along the strike of the structure than when measured perpendicularly to the structure. The difference may be of the order of 5–15%. In the case of a horizontal bedding with alternating hard and soft layers, the velocities will differ in vertical and horizontal directions. Such velocity discrepancies can be revealed by comparing the velocities obtained in the horizontal direction with those measured in drill holes by uphole or downhole recordings. The term uphole refers to a technique when the shots are fired in the drill hole and the arrival times are recorded on the ground surface. In downhole measurements the positions of the impacts and the detectors are reversed. It is likely that there is a certain vertical velocity alternation in overburden and in sedimentary rock formations, but possible errors caused by this are probably included in the normal uncertainty of the method discussed in Chapter 1. There are, however, more serious cases with reversed velocity relations, described in Section 3.7.

For our purpose the seismic waves may be considered to be transmitted according to the same laws as for ray optics. The three basic concepts, namely Huygens' principle, Snell's law and Fermat's principle, are vital for an understanding of wave propagation.

2.5 HUYGENS' PRINCIPLE

The principle states that each point on a wave surface acts as a source for an expanding spherical wave and after a certain time lapse the envelope of all the wavelets defines the new wave front. The principle is used for the construction of wave fronts, provided that the location of a wave front at a particular time and the velocity in the medium are known. Sometimes, however, it is more convenient to study the propagation directions in the form of raypaths instead of wave fronts. The raypaths are perpendicular to the wave fronts in isotropic media. It should be noted that the wave fronts are physical realities while raypaths are simplified, sometimes not completely true, constructions.

2.6 SNELL'S LAW

The law is a consequence of Fermat's principle, which states that an elastic disturbance travels from one point to another along a path that requires the minimum amount of time. The statement implies that the shortest travel time does not necessarily refer to a straight line between the points, if the points are located in different media with different physical properties.

When an incident wave strikes an interface separating two media, each point on the interface becomes the source of a hemispherical wave travelling into the second medium with the velocity in that medium. In Fig. 2.6 the wave through the upper layer impinges obliquely on the interface between the two layers having the velocities V_1 and V_2. It is assumed that V_1 is less than V_2 and that the wave front AB in the upper medium is to be regarded as plane. The incident wave gives rise to two new waves on reaching the interface. The energy is partly reflected back into the upper medium and partly refracted into the lower medium. The reflected wave fronts, the raypaths of which are shown by dashed lines, make the same angle with the interface as the incident wave fronts and consequently the raypaths of the incident and reflected waves make the same angle i with the normal to the interface. In the lower medium,

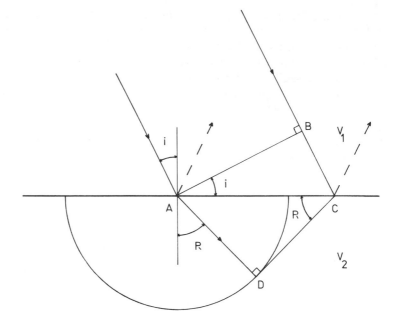

Fig. 2.6   Snell's law ($V_1 < V_2$).

the waves change direction since the properties of the two media are different, i.e. the waves are refracted.

If we study the refraction case it can be seen in Fig. 2.6 that when the wave front strikes the interface at A its position on another ray is B in the upper medium. During the time t that the wave in the upper medium travels from B to C where BC is equal to $V_1 t$, the point A acts as a source for a hemispherical wave front propagating in the lower medium, the radius of the hemisphere being $V_2 t$. From C a tangent is drawn to the semicircle. The tangent is the envelope of the wavelets emanating during time t from each point of the interface between A and C.

From the quadrilateral ABCD

$$\sin i = \frac{BC}{AC}$$

and

$$\sin R = \frac{AD}{AC}$$

so that

$$\frac{\sin i}{\sin R} = \frac{BC}{AD} = \frac{V_1 t}{V_2 t} = \frac{V_1}{V_2} \tag{2.13}$$

This equation is known as Snell's law or the law of refraction. The angle R is called the angle of refraction and i the angle of incidence. Since V_2 in this case is greater than V_1, R is greater than i. When i increases there is a unique case where the angle of refraction R is 90° and $\sin R = 1$.

Therefore, in this particular case

$$\sin i = \frac{V_1}{V_2} = \sin i_{12} \tag{2.14}$$

The angle i_{12} is called the critical angle of incidence. For incidence angles greater than i_{12}, the energy is totally reflected into the upper layer. Snell's law cannot be satisfied since $\sin R$ cannot exceed unity.

The analysis above is valid as long as the velocity in each underlying layer in a sequence is higher than the velocity in the layer above. For a case where V_2 is less than V_1, the angle of refraction R is less than the angle of incidence i. This case is shown in Fig. 2.7. When the wave in the upper medium travels the distance BC, the wave in the lower medium travels the shorter distance AD. The rays in the V_2-layer are deflected downwards.

When in Fig. 2.6 the incidence angle is i_{12}, the travel time for BC in the upper medium is equal to the travel time for AC in the lower medium. The wave front in the lower medium is in this case perpendicular to the interface

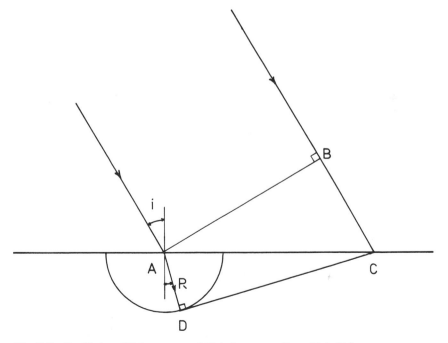

Fig. 2.7   Snell's law (V_1 in upper and V_2 in lower medium, $V_1 > V_2$).

between the two media in the vicinity of the interface. A wave travelling along the top of the second medium generates a secondary wave in the upper medium due to oscillating stress at the interface. This is shown in Fig. 2.8 where V_2 is assumed to be greater than V_1. When a wave in the lower medium travels the distance EG, the hemisphere in the upper medium has attained a radius EF. The line GF is the envelope of all the wavelets emanating from points along EG, i.e. it is the wave front in the upper medium. The distances

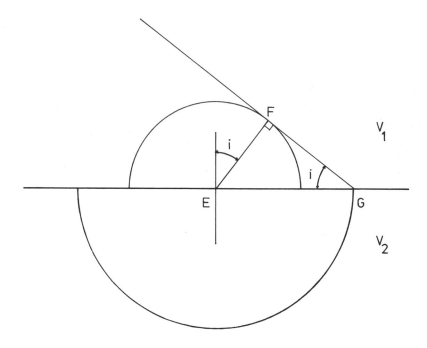

Fig. 2.8   Critical angle.

EG and EF are V_2t and V_1t respectively. The angle i between the wave front and the interface is given by the relation

$$\sin i = \frac{EF}{EG} = \frac{V_1t}{V_2t} = \frac{V_1}{V_2} \tag{2.15}$$

The angle i is thus equal to the aforementioned critical angle i_{12} and consequently the raypaths make the angle i_{12} with the normal to the interface. The waves are said to be critically refracted into the upper medium. The analysis has demonstrated that the refraction principle is based on wave fronts making the critical angle i_{12} with the interface between the media and the corresponding raypaths making the angle i_{12} with the normal to the interface.

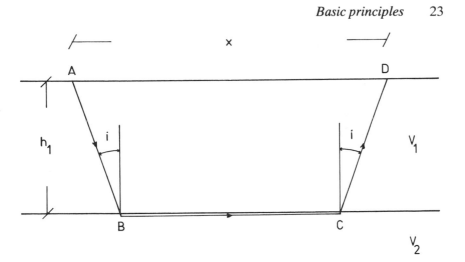

Fig. 2.9 Critical refraction.

But a question remains to be answered. Does a trajectory based on critical angles give the shortest travel time?

Referring to Fig. 2.9, the statement is that the least travel time from A to D via the second layer is obtained when the incidence angle i is equal to the critical angle i_{12}. The total travel time T_{AD} refers to three legs, where AB $=$ CD $= h_1/\cos i$ and BC $= x - 2h_1 \tan i$. It is assumed that V_1 is less than V_2.

Thus

$$T_{AD} = \frac{2h_1}{V_1 \cos i} + \frac{x - 2h_1 \tan i}{V_2}$$

The derivative of T_{AD} with respect to i is

$$\frac{dT_{AD}}{di} = \frac{2h_1 \sin i}{V_1 \cos^2 i} - \frac{2h_1}{V_2 \cos^2 i}$$

If this is equated to zero we get

$$\sin i = \frac{V_1}{V_2} = \sin i_{12}$$

Therefore, the travel time from A to D via the second layer is either a minimum or a maximum, when the slant raypaths through the V_1-layer make the angle i_{12} with the normal to the interface. The second derivative confirms that the time is a minimum.

For multilayer cases the raypaths of least time follow the same laws as for a

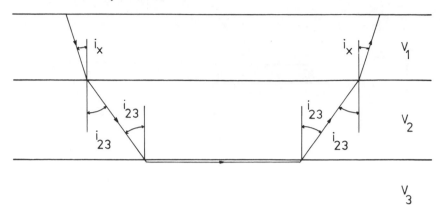

Fig. 2.10   Multilayer refraction.

two-layer case, but the conditions are somewhat more complicated. In Fig. 2.10 the critical angle i_{23} is given by the relation $\sin i_{23} = V_2/V_3$. The unknown incidence angle i_x is expressed according to Snell's law by

$$\frac{\sin i_x}{\sin i_{23}} = \frac{V_1}{V_2}$$

Replacing $\sin i_{23}$ by V_2/V_3, we obtain

$$\sin i_x = \frac{V_1 V_2}{V_2 V_3} = \frac{V_1}{V_3} \tag{2.16}$$

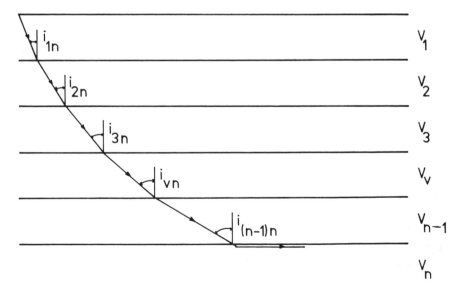

Fig. 2.11   Raypaths and critical angles in multilayer refraction.

The incidence angle in the upper layer is designated i_{13} and is obtained from the equation $\sin i_{13} = V_1/V_3$. The general expression for the angle of incidence for multilayer cases is $\sin i_{vn} = V_v/V_n$. The equation for the critical angle is $\sin i_{(n-1)n} = V_{n-1}/V_n$. The raypaths and angles of incidence are shown in Fig. 2.11.

2.7 DIFFRACTION

Besides direct and refracted waves, there is a third important type of wave, namely the diffracted wave, which spreads out over spheres. These waves are mainly due to edges in the refracting interfaces, associated with faults, prominent peaks or velocity changes in the refractor.

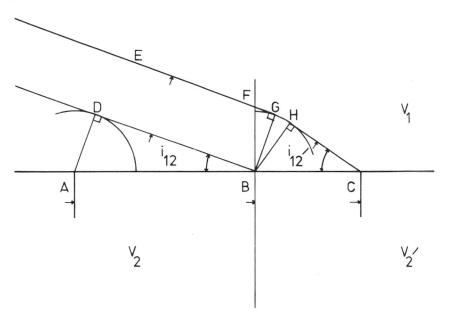

Fig. 2.12 Origin of a diffraction zone.

A case with a diffraction zone caused by a velocity change in the refractor is shown in Fig. 2.12. The velocities V_2 and $V_{2'}$ are separated by a vertical boundary, forming an edge at B. It is assumed that $V_1 < V_{2'} < V_2$. When the wave has passed from A to B in the second medium, the resulting refracted wave front (head wave) in the V_1-layer has reached the position given by the line DB. The edge at B acts as a source of waves radiating into the three media with the respective velocities. In this case we consider only the wave propagation towards the right from B. Since the velocity boundary is perpendicular to the raypaths in the lower medium, the wave continues to the right of B without being refracted. When this wave is at C, the wave returning from B into the upper layer forms a semicircle represented by the arc GH. The head wave front emanating from the $V_{2'}$-layer is given by the line HC while the line EG, parallel to DB, is the corresponding head wave front from the V_2-layer at that moment. The wave front HC cannot exist to the left of H and the other wave front EG to the right of G, since wave fronts have to be perpendicular to the raypaths. Therefore, within the sector HBG the waves radiate as spheres forming a diffraction zone. With increasing thickness of the upper layer the zone widens. The angle of the diffraction zone is $i_{12'} - i_{12}$, since the angles

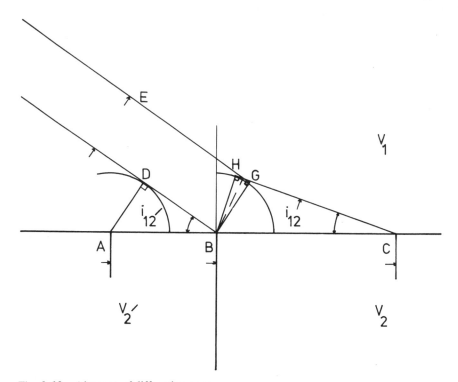

Fig. 2.13   Absence of diffraction zone.

FBG and FBH are i_{12} and $i_{12'}$ respectively. A wave propagation from a higher to a lower velocity generates a diffraction zone in the overlying layer. The reverse condition, namely propagation from lower to higher velocity, does not give rise to diffraction. This case is exemplified in Fig. 2.13. The wave propagation is also here from the left to the right. The course of events is similar to that in Fig. 2.12, but for the wave fronts EG and HC the contact points H and G have changed position on the circle with B as centre. The wave front contact, dashed line, makes the angle $(i_{12'} + i_{12})/2$ with the normal to the interface, in this case the vertical through B.

A velocity boundary within a refractor can give rise to diffraction zones not

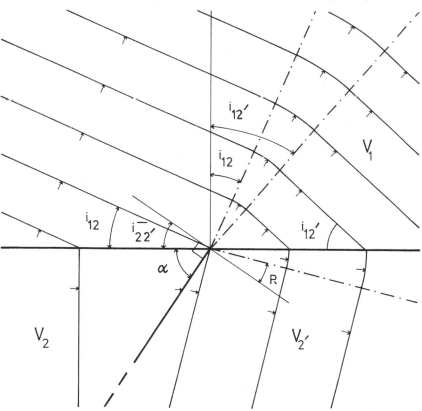

Fig. 2.14   Diffraction within a refractor (after Sjögren, 1979).

only in the overlying layer but also inside the refractor itself. In Figs 2.14 and 2.15 there is a velocity variation in the second layer and the boundary separating V_2 and $V_{2'}$ is not perpendicular to the interface between the upper and lower layers. The model studies assume that $V_1 < V_{2'} < V_2$. Since the raypaths for the incident waves are assumed to be parallel to the interface, the wave fronts make an angle with the velocity boundary in the bottom layer. Therefore, when the waves pass the boundary they are refracted.

In Fig. 2.14 the waves, impinging obliquely on the boundary between the V_2 and $V_{2'}$ sections, are refracted downwards, and along the $V_{2'}$ section only a part of the energy spreads outwards within a diffraction zone. The boundary between the refracted and diffracted waves is shown by a dash-dotted line. In the upper medium the waves are critically refracted except for those in the diffraction zone. As the wave propagation is from a higher to a lower velocity, the incident angle $i_{\overline{22'}}$ is greater than the refraction angle R. The symbol $i_{\overline{22'}}$ indicates a wave propagation from V_2 to $V_{2'}$ when the angle of incidence is

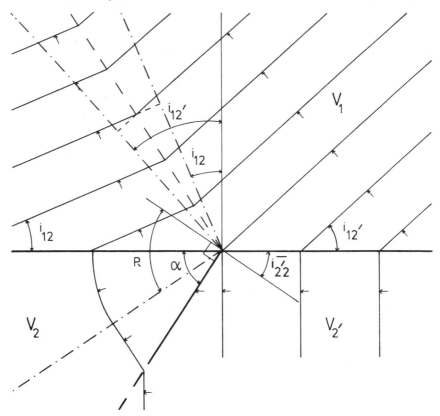

Fig. 2.15 Diffraction within a refractor (after Sjögren, 1979).

non-critical. Snell's law gives the relation $\sin R = V_{2'}\cos\alpha/V_2$ since $i_{\overline{22'}}$ is $90° - \alpha$.

Figure 2.15 shows the same structure as Fig. 2.14, but the wave propagation is reversed. Once again the main part of the energy is deflected downwards and at the surface of the V_2 section there is a diffraction zone. The angle of refraction R is greater than the angle of incidence $i_{\overline{2'2}}$ since the waves propagate from a lower to a higher velocity section. According to Snell's law the relation between the angles is obtained from $\sin R = V_2\cos\alpha/V_{2'}$, since the angle of incidence $i_{\overline{2'2}}$ is equal to $90° - \alpha$. The dashed line in the V_1-layer indicates the wave front contact and the dash-dotted lines the limits outside which the respective waves cannot be generated. The wave front contact makes the angle $(i_{12} + i_{12'})/2$ with the normal to the layer interface.

2.8   NON-CRITICAL REFRACTION

The method is based on waves critically refracted into the upper layers, but in reality we have to reckon with a lot of non-critical refraction. This type of refraction is generated when the wave fronts are not normal to the surface of the refractor, or, in other words, when the raypaths make an angle with the refractor surface. In the two-layer case of Fig. 2.16 there is a dip change at B. The arrows in the figure denote the direction of the wave propagation. To the left of B the waves are critically refracted into the V_1-layer, as the wave fronts are perpendicular to the refractor. On the other hand, to the right of B, the wave fronts make a non-right angle with the refractor surface. The angle R is

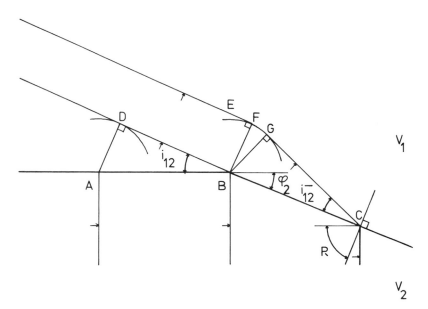

Fig. 2.16 Non-critical refraction due to dip.

equal to $90° - \varphi_2$. The corresponding non-critical i_{12}^- is obtained according to Snell's law from the equation

$$\frac{\sin i_{12}^-}{\sin (90 - \varphi_2)} = \frac{V_1}{V_2}$$

and

$$\sin i_{12}^- = \frac{V_1 \cos \varphi_2}{V_2}$$

In Fig. 2.16 the angle EBF $= i_{12}$ and the angle EBG $= i_{12}^- + \varphi_2$. The non-critical angle i_{12}^- is less than the critical angle i_{12}. The dip change at B has given rise to a diffraction zone in the V_1-layer, namely the sector GBF.

Velocity boundaries within a refractor can also cause non-critical refraction. This is illustrated in Figs 2.17 and 2.18. The figures show the same

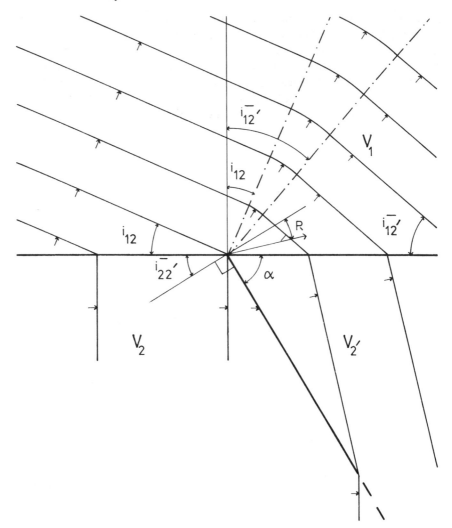

Fig. 2.17 Non-critical refraction due to velocity boundary (after Sjögren, 1979).

structure in principle as in Figs 2.14 and 2.15, except for a change in the dip of the velocity boundary within the second layer.

In Fig. 2.17 the angle of incidence $\overline{i_{22'}}$ is equal to $90° - \alpha$. Snell's law gives $\sin R = V_{2'} \cos\alpha / V_2$. After passing the velocity boundary the waves are bent upwards and the rays make the angle $90° - (\alpha + R)$ with the refractor surface. The sine of the non-critical angle $\overline{i_{12'}}$ in the V_1-layer is equal to $V_1 \sin(\alpha + R)/V_{2'}$. The recorded velocity from the $V_{2'}$-layer will be overestimated because the real travel paths through the layer are shorter than the corresponding

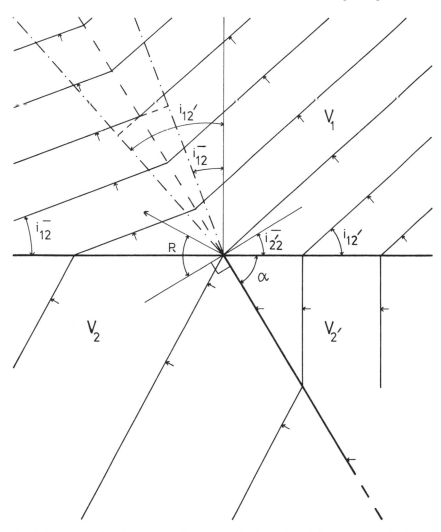

Fig. 2.18 Non-critical refraction due to velocity boundary (after Sjögren, 1979).

distances along the surface of the layer. Wave propagation in the reverse direction, Fig. 2.18, will also lead to an overestimate of the refractor velocity, in this case from the V_2 section. The refraction angle R is given by $\sin R = V_2 \cos \alpha / V_{2'}$. The angle of incidence $i_{2'2}^- = 90° - \alpha$. The non-critical angle i_{12}^- is expressed (by Snell's law) as $\sin i_{12}^- = V_1 \sin (\alpha + R)/V_2$. The rays within the V_2 section make the angle $R + \alpha - 90°$ with the refractor surface.

3

Depth formulae

The two-layer case shown in Fig. 3.1 will be used to explain some fundamental principles. From the impact point B, the waves spread out in accordance with the laws given in Chapter 2 into the two media composing the subsurface.

In the vicinity of point B, the waves travel in the upper layer with velocity V_1 and on reaching the second layer they propagate with the higher velocity V_2. The waves in the second layer generate waves in the upper layer because of an oscillating stress at the interface. These latter waves return to the surface as plane waves making the angle i_{12} with the interface. The corresponding raypaths, shown in the cross-section (b), make the angle i_{12} with the normal to the interface (see Section 2.6).

At large distances from B the waves that travel along the longer but faster path in the second medium will overtake those that follow the ground surface. Therefore, there must exist a point along the surface where the times for the arrival of the direct waves and refracted waves are equal.

If the first arrivals of the elastic waves are recorded by detectors planted in the ground, the times from the impact instant to the detectors can be plotted on a time–distance graph as shown by the dots in the upper part of Fig. 3.1. The slopes of the lines obtained by connecting these dots yield the reciprocal of the velocities, namely $1/V$. Therefore, the lower the velocity, the steeper the slope of the time–distance line. The intersection, 'break point', between the two velocity lines is obviously the point where the times are equal. The distance between the impact point B and the break point is called the critical distance. The break point corresponds to the emergence of the wave front contact at the ground surface. The contact representing the locus of points for which the times are equal within the V_1-layer, is shown by a broken line in section (a). However, the wave fronts exist on both sides of the wave front contact and may be recorded at the ground surface as second arrivals. On the time–distance graph second arrivals are indicated by dashed lines.

For calculating the depth at an impact point two different approaches are available using either the intercept or the critical distance. The intercept time,

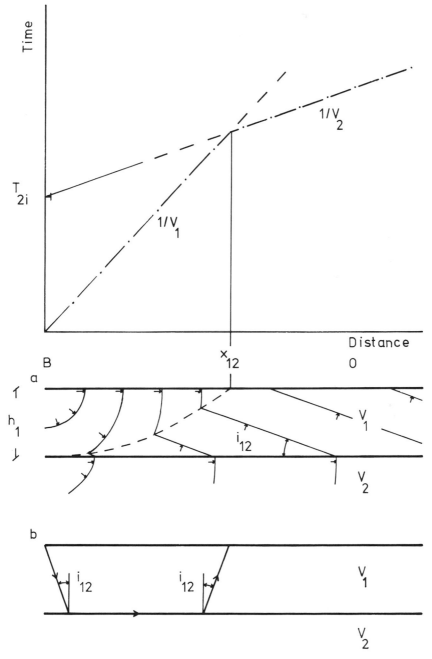

Fig. 3.1 Fundamental principle of refraction shooting.

marked T_{2i} on the graph, is the intersection between the prolongation of the time–distance segment corresponding to the second medium and the time axis through the impact point. In order to obtain the intercept time we need the time–distance–depth equation for the segment corresponding to the second layer in Fig. 3.1. In the usual mathematical manner, the equation can be used to obtain the intercept time by assuming that the distance is equal to zero. The critical distance is obtained by equating the times for the first and second layers. Developments of the concepts are given below.

3.1   TWO-LAYER CASE

The algebraic notation in what follows refers to Fig. 3.1, which shows a simple case with constant velocities and horizontal plane interfaces. The critical angle i_{12} is given by the relation $\sin i_{12} = V_1/V_2$. The depth to the second layer is h_1.

3.1.1   Intercept time

The equation for the arrival time T_1 of the direct surface waves is

$$T_1 = \frac{x}{V_1} \tag{3.1}$$

which is the equation for a straight line through the impact point with slope $1/V_1$.

For the arrival time T_2 of the refracted waves we have

$$T_2 = \frac{2h_1}{V_1 \cos i_{12}} + \frac{x - 2h_1 \tan i_{12}}{V_2} \tag{3.2}$$

$$= \frac{2h_1}{V_1 \cos i_{12}} - \frac{2h_1 \sin i_{12} \sin i_{12}}{V_1 \cos i_{12}} + \frac{x}{V_2}$$

$$= \frac{2h_1}{V_1 \cos i_{12}} - \frac{2h_1(1 - \cos^2 i_{12})}{V_1 \cos i_{12}} + \frac{x}{V_2}$$

$$= \frac{x}{V_2} + \frac{2h_1 \cos i_{12}}{V_1} \tag{3.3}$$

which is the equation of a straight line with slope $1/V_2$ and an intercept on the time axis through the impact point (i.e. the time for $x = 0$) equal to

$$T_{2i} = \frac{2h_1 \cos i_{12}}{V_1} \tag{3.4}$$

From this we get

$$h_1 = \frac{T_{2i}V_1}{2\cos i_{12}} \tag{3.5}$$

The formula can also be expressed in terms of the velocities V_1 and V_2 as

$$h_1 = \frac{T_{2i}V_1V_2}{2\sqrt{(V_2^2 - V_1^2)}} \tag{3.6}$$

since

$$\cos i_{12} = \sqrt{(1 - V_1^2/V_2^2)}$$

The intercept time is the total time T_2 minus the time x/V_2, x being the distance BO. The real travel path in the V_2-layer is, however, not equal to x but to $x - 2h_1 \tan i_{12}$ because of the slant raypaths in the upper medium. We have here the concept of delay time introduced by Gardner in 1939. A delay time is defined as the travel time for any slant raypath between the ground surface (reference level) and a refractor *minus* the time required to travel the horizontal projection of the raypath at the velocity of the refractor.

For the mathematical expression of the delay time concept, I refer to Fig. 3.2. Assume that A is the impact point and E the receiving station. The velocities are constant but the interface depths at A and E are different. According to definition, the delay time δ at A is the time for the slant raypath AC minus the time for the path BC.

Therefore

$$\delta = \frac{h_A}{V_1 \cos i_{12}} - \frac{h_A \tan i_{12}}{V_2} \tag{3.7}$$

Since $V_2 = V_1/\sin i_{12}$, we obtain

$$\delta = \frac{h_A}{V_1 \cos i_{12}} - \frac{h_A \sin^2 i_{12}}{V_1 \cos i_{12}}$$

$$= \frac{h_A}{V_1 \cos i_{12}} - \frac{h_A(1 - \cos^2 i_{12})}{V_1 \cos i_{12}}$$

$$= \frac{h_A \cos i_{12}}{V_1} \tag{3.8}$$

As can be seen from Fig. 3.2, the time along the path AD at velocity V_1 is equal to the delay time.

On comparing the intercept time in Equation (3.4) and the delay time in

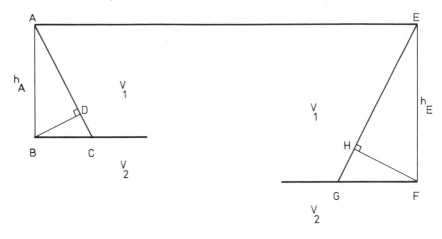

Fig. 3.2   Delay time concept.

Equation (3.8), we see that the intercept time is composed of two delay times, one at the impact point and another at the receiving station. The delay times are identical since the layers in Fig. 3.1 are horizontal and the velocities are constant. On the other hand, in Fig. 3.2 it has been assumed that the interface depth is not constant and therefore the delay times composing the intercept time are not equal. One of the main problems in refraction depth determination is to partition the intercept time into one delay time at the impact point and another at the detector position. This question will be discussed further in Chapter 4.

3.1.2   Critical distance

The intersection of the two time–distance lines in Fig. 3.1 is the critical distance x_{12} at which $T_1 = T_2$.

Thus

$$\frac{2h_1}{V_1\cos i_{12}} + \frac{x_{12} - 2h_1\tan i_{12}}{V_2} = \frac{x_{12}}{V_1} \tag{3.9}$$

$$\frac{2h_1}{V_1\cos i_{12}} - \frac{2h_1(1-\cos^2 i_{12})}{V_1\cos i_{12}} = x_{12}\left(\frac{1}{V_1} - \frac{1}{V_2}\right)$$

Using the relation $V_1/V_2 = \sin i_{12}$

$$h_1 = \frac{x_{12}(1-\sin i_{12})}{2\cos i_{12}} \tag{3.10}$$

In terms of the velocities instead of the critical angle i_{12}, we have

$$h_1 = \frac{x_{12}}{2}\sqrt{\left(\frac{V_2 - V_1}{V_2 + V_1}\right)} \tag{3.11}$$

3.2   THREE-LAYER CASE

Figure 3.3 shows a case with two layers (velocities V_1 and V_2) overlying the bottom refractor where the velocity is V_3. Two of the angles of incidence are critical, viz. i_{12} and i_{23}. The derivation of the angle i_{13} is given in Equation (2.16). The thickness of the V_1-layer is calculated according to the formula for a two-layer case, Equations (3.5) and (3.6) or (3.10) and (3.11). The time–distance–depths relations for the second layer are obtained in a similar manner as for the two-layer case. However, we have to consider the slant raypaths through the first as well as the second layer and their influence on the raypath along the top of the third layer.

3.2.1   Intercept time

The equation for the time of arrival T_3 for the wave from B to O via the third layer is

$$T_3 = \frac{2h_1}{V_1\cos i_{13}} + \frac{2h_2}{V_2\cos i_{23}} + \frac{x - 2h_1\tan i_{13} - 2h_2\tan i_{23}}{V_3} \tag{3.12}$$

$$= \frac{2h_1}{V_1\cos i_{13}} + \frac{2h_2}{V_2\cos i_{23}} - \frac{2h_1\sin i_{13}\sin i_{13}}{V_1\cos i_{13}} - \frac{2h_2\sin i_{23}\sin i_{23}}{V_2\cos i_{23}} + \frac{x}{V_3}$$

$$= \frac{2h_1}{V_1\cos i_{13}} + \frac{2h_2}{V_2\cos i_{23}} - \frac{2h_1}{V_1\cos i_{13}} + \frac{2h_1\cos^2 i_{13}}{V_1\cos i_{13}} - \frac{2h_2}{V_2\cos i_{23}} +$$

$$+ \frac{2h_2\cos^2 i_{23}}{V_2\cos i_{23}} + \frac{x}{V_3}$$

$$= \frac{2h_1\cos i_{13}}{V_1} + \frac{2h_2\cos i_{23}}{V_2} + \frac{x}{V_3} \tag{3.13}$$

The equation for T_3 represents a straight line with slope $1/V_3$ and (when $x = 0$) an intercept time

$$T_{3i} = \frac{2h_1\cos i_{13}}{V_1} + \frac{2h_2\cos i_{23}}{V_2} \tag{3.14}$$

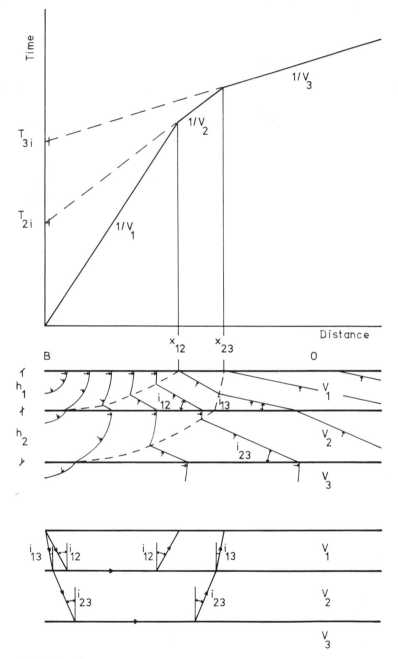

Fig. 3.3 Three-layer case.

Solving for h_2, we obtain

$$h_2 = \frac{T_{3i}V_2}{2\cos i_{23}} - \frac{V_2 h_1 \cos i_{13}}{V_1 \cos i_{23}} \tag{3.15}$$

or, when V_2/V_1 is replaced by $1/\sin i_{12}$ in the second term on the right-hand side of the equation

$$h_2 = \frac{T_{3i}V_2}{2\cos i_{23}} - \frac{h_1 \cos i_{13}}{\cos i_{23} \sin i_{12}} \tag{3.16}$$

3.2.2 Critical distance

The intersection between the straight lines for T_2 in Equation (3.3) and T_3 in Equation (3.13) defines the distance x_{23}.

Putting $T_2 = T_3$, we obtain

$$\frac{x_{23}}{V_3} + \frac{2h_1 \cos i_{13}}{V_1} + \frac{2h_2 \cos i_{23}}{V_2} = \frac{x_{23}}{V_2} + \frac{2h_1 \cos i_{12}}{V_1} \tag{3.17}$$

$$\frac{2h_2 \cos i_{23}}{V_2} = \frac{x_{23}(V_3 - V_2)}{V_2 V_3} - \frac{2h_1}{V_1}(\cos i_{13} - \cos i_{12})$$

Solving for h_2, we obtain

$$h_2 = \frac{x_{23}(V_3 - V_2)}{2V_3 \cos i_{23}} - \frac{h_1 V_2}{V_1 \cos i_{23}}(\cos i_{13} - \cos i_{12})$$

$$= \frac{x_{23}(V_3 - V_2)}{2\sqrt{(V_3^2 - V_2^2)}} - \frac{h_1(\cos i_{13} - \cos i_{12})}{\cos i_{23} \sin i_{12}}$$

$$= \frac{x_{23}}{2}\sqrt{\left(\frac{V_3 - V_2}{V_3 + V_2}\right)} - \frac{h_1(\cos i_{13} - \cos i_{12})}{\cos i_{23} \sin i_{12}} \tag{3.18}$$

or, in terms of angles

$$h_2 = \frac{x_{23}(1 - \sin i_{23})}{2\cos i_{23}} - \frac{h_1(\cos i_{13} - \cos i_{12})}{\cos i_{23} \sin i_{12}} \tag{3.19}$$

3.3 MULTILAYER CASES

The formulae above can be extended to any number of layers as long as the velocity in a layer is higher than that in the layer above. Another condition is that the layers have to be represented in the recorded arrival times. This will be discussed later on in Section 3.7. Depth equations for an arbitrary number of velocity layers are given below.

3.3.1 Intercept times

$$h_{(n-1)} = \frac{T_{ni} V_{(n-1)}}{2\cos i_{(n-1)n}} - \frac{V_{(n-1)}}{\cos i_{(n-1)n}} \sum_{v=1}^{v=n-2} \frac{h_v \cos i_{vn}}{V_v} \qquad (3.20)$$

or

$$h_{(n-1)} = \frac{T_{ni} V_n V_{(n-1)}}{2\sqrt{(V_n^2 - V_{(n-1)}^2)}} - \frac{V_n V_{(n-1)}}{\sqrt{(V_n^2 - V_{(n-1)}^2)}} \sum_{v=1}^{v=n-2} h_v \sqrt{\left(\frac{1}{V_v^2} - \frac{1}{V_n^2}\right)} \qquad (3.21)$$

3.3.2 Critical distances

$$h_{(n-1)} = x_{(n-1)n} \frac{1 - \sin i_{(n-1)n}}{2\cos i_{(n-1)n}} - \sum_{v=1}^{v=n-2} h_v \frac{\cos i_{vn} - \cos i_{v(n-1)}}{\cos i_{(n-1)n} \sin i_{v(n-1)}} \qquad (3.22)$$

or

$$h_{(n-1)} = \frac{x_{(n-1)n}}{2} \sqrt{\left(\frac{V_n - V_{(n-1)}}{V_n + V_{(n-1)}}\right)} - \sum_{v=1}^{v=n-2} h_v \frac{\cos i_{vn} - \cos i_{v(n-1)}}{\cos i_{(n-1)n} \sin i_{v(n-1)}} \qquad (3.23)$$

In the above equations T_{ni} is the intercept made by the nth line segment of the travel time curve on the time axis, $i_{(n-1)n}$ is the angle of incidence between the $(n-1)$th and the nth layer, V_n is the velocity in the nth layer, and $x_{(n-1)n}$ is the intersection (critical distance) of the $(n-1)$th and nth velocity line segments.

The depth calculations can easily be carried out on pocket calculators or by nomograms, using either formula.

3.4 SLOPING LAYERS

In Fig. 3.4 the refractor surface is sloping, making an angle φ_2 with the horizontal ground surface. The terms z_d and z_u designate the perpendicular distances between the energy sources at A and D and the refractor. The vertical depths are denoted by h_d and h_u. We now assume that rays such as ABCD, making the critical angle i_{12} ($\sin i_{12} = V_1/V_2$) with the normal to the refractor, take the shortest time from A to D and are therefore 'first arrivals'. The validity of this assumption will be proved later.

From Fig. 3.4 we have (with AD = x)

$$AB = CI = z_d/\cos i_{12}$$

$$DK = x\sin\varphi_2$$

$$ID = x\sin\varphi_2/\cos i_{12}$$

$$CD = (z_d + x\sin\varphi_2)/\cos i_{12}$$

$$AK = EG = x\cos\varphi_2$$

$$EB = z_d\tan i_{12}$$

and

$$CG = (z_d + x\sin\varphi_2)\tan i_{12}$$

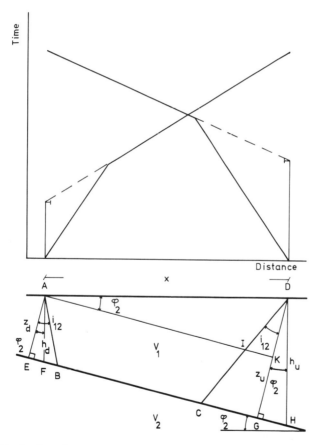

Fig. 3.4 Sloping interface.

If we assume point A to be the energy source and D the detector station, the time from A to D for the ray ABCD, i.e. the 'down-dip' time T_{2d}, is

$$T_{2d} = \frac{AB}{V_1} + \frac{BC}{V_2} + \frac{CD}{V_1}$$

$$= \frac{z_d}{V_1 \cos i_{12}} + \frac{x \cos \varphi_2 - z_d \tan i_{12} - (z_d + x \sin \varphi_2) \tan i_{12}}{V_2} + \frac{z_d + x \sin \varphi_2}{V_1 \cos i_{12}}$$

$$(3.24)$$

$$= \frac{2z_d}{V_1 \cos i_{12}} - \frac{2z_d \tan i_{12} \sin i_{12}}{V_1} + \frac{x \cos \varphi_2 \sin i_{12}}{V_1} - \frac{x \sin \varphi_2 \tan i_{12} \sin i_{12}}{V_1}$$

$$+ \frac{x \sin \varphi_2}{V_1 \cos i_{12}}$$

$$= \frac{2z_d}{V_1 \cos i_{12}} (1 - \sin^2 i_{12}) + \frac{x}{V_1} \left(\cos \varphi_2 \sin i_{12} - \frac{\sin \varphi_2 \sin^2 i_{12}}{\cos i_{12}} + \frac{\sin \varphi_2}{\cos i_{12}} \right)$$

$$= \frac{2z_d \cos i_{12}}{V_1} + \frac{x}{V_1} (\cos \varphi_2 \sin i_{12} + \cos i_{12} \sin \varphi_2)$$

$$= \frac{2z_d \cos i_{12}}{V_1} + \frac{x}{V_1} \sin (i_{12} + \varphi_2) \qquad (3.25)$$

This equation represents a straight line with slope $\sin (i_{12} + \varphi_2)/V_1$, and when $x = 0$, the intercept is equal to $2z_d \cos i_{12}/V_1$. The apparent velocity $V_1/\sin (i_{12} + \varphi_2)$ is equal to $V_2 \sin i_{12}/\sin (i_{12} + \varphi_2)$ and is smaller than the true velocity V_2 since $\sin i_{12}/\sin (i_{12} + \varphi_2)$ is less than unity.

Similarly, we obtain the time–distance equation for the up-dip recording by replacing z_d by $(z_u - x \sin \varphi_2)$ in Equation (3.25). A more direct way is to replace z_d by z_u and φ_2 by $-\varphi_2$, considering the dip to be negative.

Hence

$$T_{2u} = \frac{2z_u \cos i_{12}}{V_1} + \frac{x}{V_1} \sin (i_{12} - \varphi_2) \qquad (3.26)$$

The apparent velocity $V_1/\sin (i_{12} - \varphi_2) = V_2 \sin i_{12}/\sin (i_{12} - \varphi_2)$ is now higher than the true velocity.

The derivations above were based on the assumption that the least time from A to D is obtained when the rays make the angle i_{12} with the refractor surface. This has to be proved.

The total time for an arbitrary raypath from A to D is obtained simply by replacing i_{12} in Equation (3.24) by an arbitrary angle of incidence i.

$$T_{2d} = \frac{2z_d}{V_1 \cos i} - \frac{2z_d \tan i}{V_2} + \frac{x \cos \varphi_2}{V_2} - \frac{x \sin \varphi_2 \tan i}{V_2} + \frac{x \sin \varphi_2}{V_1 \cos i} \tag{3.27}$$

The derivative of T_{2d} with respect to i is

$$\frac{dT_{2d}}{di} = \frac{2z_d \sin i}{V_1 \cos^2 i} - \frac{2z_d}{V_2 \cos^2 i} - \frac{x \sin \varphi_2}{V_2 \cos^2 i} + \frac{x \sin \varphi_2 \sin i}{V_1 \cos^2 i} = 0$$

Rearranging terms, we obtain

$$\frac{\sin i}{V_1} (2z_d + x \sin \varphi_2) = \frac{1}{V_2} (2z_d + x \sin \varphi_2)$$

so that

$$\sin i = \frac{V_1}{V_2} \tag{3.28}$$

which is the velocity relation for the critical angle i_{12}. The second derivative proves that the condition is for minimum time.

The slopes of the refractor velocity segments in Fig. 3.4 are $\sin(i_{12} + \varphi_2)/V_1$ and $\sin(i_{12} - \varphi_2)/V_1$. The corresponding reciprocals are the apparent velocities designated by V_{2d} and V_{2u} respectively. Hence

$$V_{2d} = V_1/\sin(i_{12} + \varphi_2) \text{ and } V_{2u} = V_1/\sin(i_{12} - \varphi_2) \tag{3.29}$$

Solving for i_{12} and φ_2, we obtain

$$i_{12} = \frac{1}{2}\left(\sin^{-1}\frac{V_1}{V_{2d}} + \sin^{-1}\frac{V_1}{V_{2u}}\right) \tag{3.30}$$

and

$$\varphi_2 = \frac{1}{2}\left(\sin^{-1}\frac{V_1}{V_{2d}} - \sin^{-1}\frac{V_1}{V_{2u}}\right) \tag{3.31}$$

The intercepts are

$$T_{2di} = 2z_d \cos i_{12}/V_1 \text{ and } T_{2ui} = 2z_u \cos i_{12}/V_1 \tag{3.32}$$

so that

$$z_d = T_{2di} V_1/2\cos i_{12} \text{ and } z_u = T_{2ui} V_1/2\cos i_{12} \tag{3.33}$$

The vertical depths h_d and h_u are obtained by dividing z_d and z_u by $\cos \varphi_2$.

If the critical distance is used, the vertical depths are given by the equations

$$h_d = x_{12} \frac{1 - \sin(i_{12} + \varphi_2)}{2\cos\varphi_2 \cos i_{12}} \quad \text{and} \quad h_u = x_{12} \frac{1 - \sin(i_{12} - \varphi_2)}{2\cos\varphi_2 \cos i_{12}} \tag{3.34}$$

The true velocity V_2 in the refractor can be derived as follows

$$\sin(i_{12} + \varphi_2) = V_1/V_{2d} \quad \text{and} \quad \sin(i_{12} - \varphi_2) = V_1/V_{2u}$$

Hence

$$\sin i_{12} \cos\varphi_2 + \cos i_{12} \sin\varphi_2 = V_1/V_{2d}$$

and

$$\sin i_{12} \cos\varphi_2 - \cos i_{12} \sin\varphi_2 = V_1/V_{2u}$$

By adding and noting that $\sin i_{12} = V_1/V_2$, we obtain

$$\frac{2\cos\varphi_2}{V_2} = \frac{V_{2u} + V_{2d}}{V_{2u} V_{2d}}$$

or

$$V_2 = 2\cos\varphi_2 \frac{V_{2u} V_{2d}}{V_{2u} + V_{2d}} \tag{3.35}$$

An estimate of the velocity can be obtained by disregarding the factor $\cos\varphi_2$. The estimate then is the harmonic mean $\bar{V}_2$ of V_{2u} and V_{2d}.

It is of interest to study in more detail the relationship between the true refractor velocity V_2 and the harmonic mean $\bar{V}_2$. In Fig. 3.5 the wave arrivals

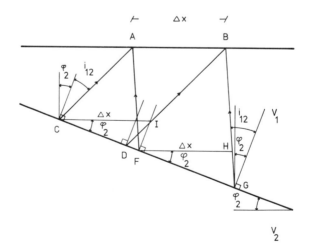

Fig. 3.5   Relationship between V_2 and $\bar{V}_2$.

recorded over the distance AB $= \Delta x$ emanate from the refractor section CD in the direct recording and from section FG in the reverse recording. In order to find the time increments over the interval AB while recording in either direction, it is only necessary to consider the raypaths CD and DI (direct) and GF and GH (reverse), since CA $=$ IB and FA $=$ HB.

In Fig. 3.5

$$CD = \Delta x \cos(i_{12} + \varphi_2)/\cos i_{12}$$

$$DI = GH = \Delta x \sin \varphi_2/\cos i_{12}$$

and

$$FG = \Delta x \cos(i_{12} - \varphi_2)/\cos i_{12}$$

The two time differences are

$$\Delta T_{\mathrm{d}} = \frac{\Delta x \cos(i_{12} + \varphi_2)}{V_2 \cos i_{12}} + \frac{\Delta x \sin \varphi_2}{V_1 \cos i_{12}} \qquad (3.36)$$

$$\Delta T_{\mathrm{u}} = \frac{\Delta x \cos(i_{12} - \varphi_2)}{V_2 \cos i_{12}} - \frac{\Delta x \sin \varphi_2}{V_1 \cos i_{12}} \qquad (3.37)$$

For the arithmetic mean ΔT of the two time increments ΔT_{d} and ΔT_{u} we obtain

$$\Delta T = \frac{1}{2}\left[\frac{\Delta x \cos(i_{12} + \varphi_2)}{V_2 \cos i_{12}} + \frac{\Delta x \cos(i_{12} - \varphi_2)}{V_2 \cos i_{12}} \right]$$

which after simplification gives

$$\Delta T = \frac{\Delta x \cos \varphi_2}{V_2}$$

The apparent velocity over the interval AB is given by

$$\frac{\Delta x}{\Delta T} = \frac{V_2}{\cos \varphi_2} = \bar{V}_2 \qquad (3.38)$$

It should be noted that in computing the arithmetic mean the second terms on the right-hand side of Equations (3.36) and (3.37) are eliminated, so that the differences in travel time through the V_1-layer do not influence the estimate of the velocity. The mean of the time increments over a fixed interval obtained by recording in the two opposite directions yields an estimate $\bar{V}_2$ of the true velocity V_2. The quantity $\bar{V}_2$ is greater than V_2 and has to be multiplied by $\cos \varphi_2$ to get V_2. The error in calculated velocity is, as pointed out above, not caused by the variation in depth but by the paths in the refractor. The distance CD is shorter and the distance FG longer than Δx and their arithmetic mean is $\Delta x \cos\varphi_2$. The overestimation of the refractor velocity is caused by the fact that the average time ΔT refers to the mean raypath of length $\Delta x \cos\varphi_2$ in the refractor but is recorded and plotted over the longer horizontal distance Δx.

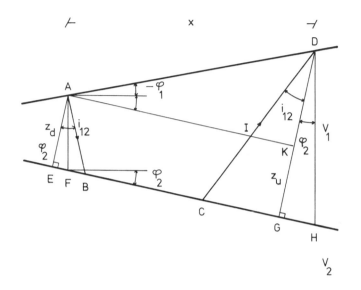

Fig. 3.6   Sloping ground surface and sloping interface.

In Fig. 3.6 both the refractor and the ground surface are inclined. In order to have the slope angles consistent, they have been attributed different signs depending on the quadrant in which they are situated.

With the situation as in the figure, we have $\angle\,DAJ = -\varphi_1 + \varphi_2$,
$ADJ = 90° - (i_{12} - \varphi_1 + \varphi_2)$, $AD = x/\cos(-\varphi_1) = x/\cos\varphi_1$,
$AB = CI = z_d/\cos i_{12}$, $DK = x\sin(\varphi_2 - \varphi_1)/\cos\varphi_1$,
$DI = x\sin(\varphi_2 - \varphi_1)/\cos\varphi_1 \cos i_{12}$, $AK = x\cos(\varphi_2 - \varphi_1)/\cos\varphi_1$,
$EB = z_d \tan i_{12}$ and $CG = [z_d + x\sin(\varphi_2 - \varphi_1)/\cos\varphi_1]\tan i_{12}$.
The total time from A to D via B and C is

$$T_d = AB/V_1 + BC/V_2 + CD/V_1$$

$$T_d = \frac{2z_d}{V_1 \cos i_{12}} + \frac{x\sin(\varphi_2 - \varphi_1)}{V_1 \cos\varphi_1 \cos i_{12}} + \frac{x\cos(\varphi_2 - \varphi_1)}{V_2 \cos\varphi_1} - \frac{2z_d \tan i_{12}}{V_2}$$

$$-\frac{x\sin(\varphi_2 - \varphi_1)\tan i_{12}}{V_2 \cos\varphi_1} \tag{3.39}$$

After simplification, the equation becomes

$$T_d = \frac{2z_d \cos i_{12}}{V_1} + \frac{x \sin[i_{12} + (\varphi_2 - \varphi_1)]}{V_1 \cos \varphi_1}$$ (3.40)

Since $z_d = z_u - x \sin(\varphi_2 - \varphi_1)/\cos \varphi_1$, the time in the other direction will be

$$T_u = \frac{2z_u \cos i_{12}}{V_1} + \frac{x \sin[i_{12} - (\varphi_2 - \varphi_1)]}{V_1 \cos \varphi_1}$$ (3.41)

When $x = 0$, we have the intercepts on the respective time-axes, namely

$$T_{2di} = 2z_d \cos i_{12}/V_1 \quad \text{and} \quad T_{2ui} = 2z_u \cos i_{12}/V_1$$ (3.42)

The apparent velocities are

$$V_{2d} = V_1 \cos \varphi_1/\sin[i_{12} + (\varphi_2 - \varphi_1)]$$ (3.43)

and

$$V_{2u} = V_1 \cos \varphi_1/\sin[i_{12} - (\varphi_2 - \varphi_1)]$$ (3.44)

The relationship between the true velocity V_2 and the harmonic mean $\bar{V}_2$ of V_{2d} and V_{2u} is

$$V_2 = \bar{V}_2 \cos(\varphi_2 - \varphi_1)/\cos \varphi_1$$ (3.45)

Because of the sloping ground surface the first segment of the time–distance curve underestimates the velocity for the V_1-layer since the actual travel path along the ground is longer than the horizontal distance traversed in the terrain and used on the time plot. The estimate of V_1 must be divided by $\cos \varphi_1$ to obtain the true velocity.

The equations above are also valid when φ_1 and φ_2 are sloping in other directions, provided that the signs of the angles are appropriate, i.e. elevation and depression angles (regarded from left to right) are given minus and plus signs respectively.

It is clear from Equation (3.45) that when the refractor surface is horizontal an inclined ground does not affect the velocity calculation. The harmonic mean $\bar{V}_2$ is equal to the true velocity V_2. On the other hand, an inclined refractor, when the ground is horizontal, always gives a harmonic-mean velocity higher than the true value and increasing with increasing dip angle φ_2. When both the ground surface and the refractor are inclined, the harmonic mean may be greater or smaller than V_2. It is always greater when the angles φ_1 and φ_2 have opposite signs. If the angles have the same sign, the velocity $\bar{V}_2$ decreases with increasing φ_1. If $2\varphi_1 = \varphi_2$ then $\bar{V}_2 = V_2$, the true velocity. For larger φ_1, $\bar{V}_2$ is lower than the true velocity.

For steep refractor slopes, infinite or negative velocities can appear, or in the case of extreme slopes the refractor velocity may not be recorded at all in

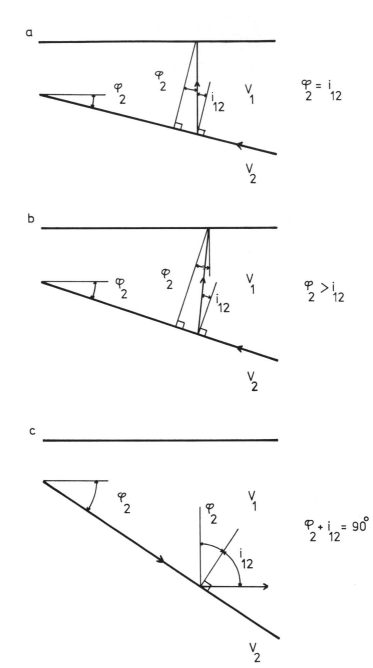

Fig. 3.7 Conditions for negative and infinite apparent velocities.

the 'down-dip' direction. These circumstances are illustrated in Fig. 3.7. The appropriate equations for the apparent, recorded velocities are $V_{2u} = V_1/\sin(i_{12} - \varphi_2)$ and $V_{2d} = V_1/\sin(i_{12} + \varphi_2)$ since only the refractor surface is inclined.

In Fig. 3.7(a) it is assumed that the angle φ_2 is equal to the angle of incidence i_{12}. The 'up-dip' apparent velocity V_{2u} is infinite. The raypaths from the refractor are vertical. Time increments caused by the wave propagation in the V_2-medium are equalled by decreases in the travel times in the over-burden. Therefore, along the traverse the arrival times remain constant and when plotted on the time–distance graph they yield a horizontal line.

With increasing φ_2, negative-velocity lines are obtained. In Fig. 3.7(b) φ_2 is greater than i_{12}. The raypaths for the waves returning to the ground surface are retrograde. The 'up-dip' equation gives a negative value for V_{2u}. The arrival times decrease with increasing distance from the impact point.

For extremely steep inclinations of the refractor, the apparent velocities in the 'down-dip' recording approach the overburden velocity. When the sum of φ_2 and i_{12} is equal to or exceeds 90°, the rays are parallel to or below the horizontal respectively. Figure 3.7(c) illustrates the case when the sum of the two angles equals 90°. Arrival times corresponding to the overburden velocity are recorded; the equation gives V_{2d} equal to V_1.

3.5   PARALLELISM AND RECIPROCITY

The travel time curves from A and D in Fig. 3.8 are the same as those given in Fig. 3.4. The time–distance graph has been supplemented by registrations from two sufficiently distant impact points, curves 1 and 2, so that the corresponding arrival times are certainly those of waves along the refractor. In Fig. 3.4 there are only registrations from one side at the beginning and at the end of the profile and velocity determination for the refractor can only be carried out for a limited section in the centre of the traverse. For a detailed and accurate interpretation it is necessary to have a complete coverage of arrival times in both directions from the refractor or, in multilayer cases, from the bottom refractor. These registrations serve many purposes, such as evalu-ation of the distribution of refractor velocities along the entire profile, analysis of the number of velocity layers and depth determination.

Travel time curves 1 and 2 are parallel to the respective refractor velocity segments for impacts from points A and D. They are due to the same layer and have identical raypaths. In practice, however, we cannot expect to get an exact parallelism between the curves due to an inevitable scattering of the plotted arrival times. When the parallelism between, for instance, the travel time curves from A and curve 1 ceases, we have the first approximation of the position of the break point. Note that the break point is apparent because of the sloping refractor surface. However, the estimate of the break point is of great help in the evaluation of the various velocity segments of the

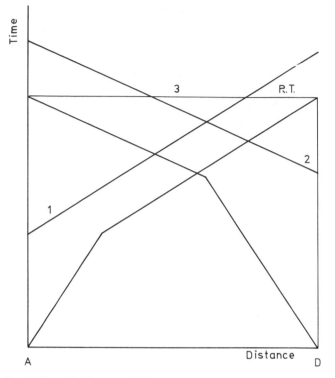

Fig. 3.8   Parallelism of refractor velocity segments.

time–distance curves, particularly when velocity differences are small or when curves are disturbed by varying geological conditions.

Another concept of importance for the analysis of the velocity picture and for the depth determination is reciprocity, i.e. the travel time from A to D via the refractor or vice versa has to be the same. The reciprocal time R.T. (the common time) is shown by the horizontal line (3) on the graph in Fig. 3.8. However, a perfect time reversal is a theoretical matter and in reality there are discrepancies.

The questions concerning parallelism and reciprocity will be discussed in more detail in Chapter 4 in connection with the various interpretation methods.

3.6   VARIOUS TRAVEL PATHS

There is an uncertainty inherent in interpreting the refraction method because of the slant raypaths through the layers overlying the refractor. The problem has been tackled in various ways. Some interpretation methods make use of raypaths converging on a common surface point while others

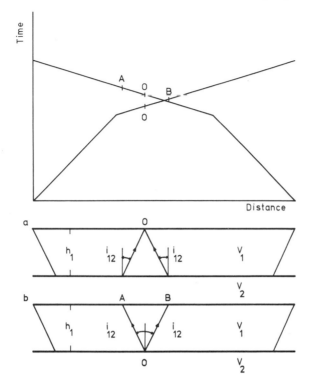

Fig. 3.9   Common surface point and common refractor point.

consider the rays emerging from a common refractor point. Both techniques have their advantages and limitations.

The raypaths used are shown in Fig. 3.9. The upper section (a) refers to the case with a common ground surface point. The critically refracted rays emerge from two points on the refractor, separated by a distance $2h_1 \tan i_{12}$, h_1 being the depth of the refractor. The corresponding arrival times have the same horizontal position, marked O on the travel time curves, but they have different origin and interpretation results are an average of the conditions at the points of emergence on the refractor.

The lower section (b) illustrates the raypath consideration with a common refractor point. The rays diverge from O on the refractor and strike the ground at A and B, the distance AB being equal to $2h_1 \tan i_{12}$. The recorded arrival times refer to the same point on the refractor but varying geological conditions at A and B may cause problems with interpretation.

The relation, AB/T, between the distance AB on the ground and the total time T for the waves along the travel path AOB is

$$2h_1 \tan i_{12} V_1 \cos i_{12} / 2h_1 = V_1 \sin i_{12}$$

This relation forms an important part of Hales' method, to be treated in Chapter 4.

3.7 HIDDEN LAYER, BLIND ZONE

In certain circumstances layers may not be detectable by the seismic refraction methods. This phenomenon is variously called in the literature the blind zone or hidden layer problem. The indetectability may be due to:

(a) Insufficient velocity contrast or insufficient thickness of an intermediate layer and
(b) Velocity inversion, i.e. the velocity in a layer is lower than that in the overlying layer

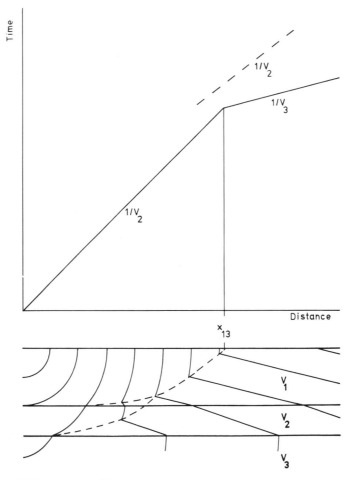

Fig. 3.10   Hidden layer problem.

In Fig. 3.10 the layer velocities are increasing with depth but the contrast between V_1 and V_2 is small. Moreover, the V_2-layer is rather thin in relation to the upper layer. A seismic record of first arrivals will only show time data from the V_1- and V_3-layers, while there will be no indications of the presence of the second layer. As can be seen in the figure, within the V_1-layer the waves from the bottom layer overtake the waves from the V_2 layer and the wave front contact is between the waves from the top and bottom layer. If second arrivals

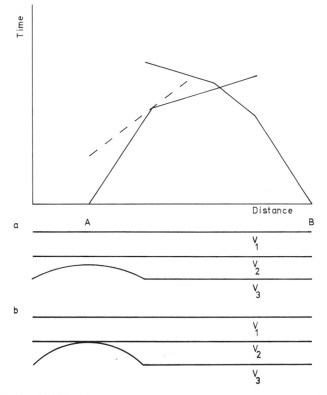

Fig. 3.11   'Semi-hidden' layer.

are available on the record, the arrival times from the V_2-layer will have the position shown by the dashed line at the top of the time–distance graph. In the absence of any knowledge of the existence of the intermediate layer, the interpretations will be carried out as for a two-layer case and the depth to the V_3-layer will be underestimated. This type of hidden layer constitutes the most serious limitation to the method since it is encountered rather often.

A common situation in seismic refraction work is that a layer may appear as what can be called 'semi-hidden', i.e. the velocity is present in some travel time curves and missing in others. This is illustrated in Figs 3.11 and 3.12.

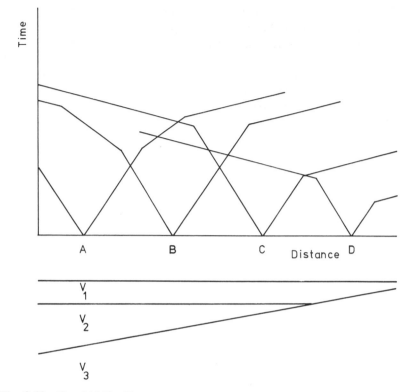

Fig. 3.12 'Semi-hidden' layer.

In Fig. 3.11 the arrival times from the impact point A refer to the V_1- and V_3-layers due to a local ridge in the V_3-layer, while at point B three velocity segments are recorded. The problem is to find out how much the thickness of the V_2-layer has decreased at A and whether it is still present as in section (a) or is completely missing as in (b). The curves from point A, based on the first arrivals, cannot give an answer to that question. The minimum depth below point A is obtained if the determination is based on the velocity segments V_1 and V_3. A maximum depth can be obtained by drawing a V_2 segment (dashed line) at the intersection between the V_1 and V_3 segments, thereby assuming that the V_2 segment has gone barely undetected. The depth calculation can be carried out using either the intercept or the critical distance formulae. If critical distances are employed, then $x_{12} = x_{23}$. It should be noted, however, that the recorded travel time curves generally yield apparent break points. If the arrival times have been corrected according to Chapter 4, the resulting new break points are to be used for the velocity segment constructions recommended above.

In Fig. 3.12 the V_2-layer thins out to the right. The presence of this layer can be observed in the direct recording from point A and in the reverse recording from point B. Maximum and minimum depths can be calculated using the technique described for the case of Fig. 3.11.

Another problem that can occur in refraction surveys is the reverse velocity relation. Refractions from an intermediate low-velocity layer cannot be obtained at the ground surface, since the waves are refracted downwards (see Fig. 2.7). The waves will return to the surface only if they have been refracted in a layer with a velocity higher than in any of the overlying layers. Velocity reversal conditions lead to an overestimate of the depths to the layers below the low-velocity layer. The existence of such layers can be ascertained by comparison with drilling results, velocity surveys in drill holes or earth resistivity measurements. Fortunately, velocity reversals do not often occur.

Sometimes the registrations can be disturbed by intermediate high-speed layers, i.e. with velocities higher than in the over- and underlying layers. Such layers are often rather thin and at a certain distance from the impact point the waves scatter and the waves from the underlying layer are recorded. Such velocity anomalies cannot directly be regarded as a disadvantage since they can be used to locate harder intermediate layers.

3.8   CONTINUOUS INCREASE IN VELOCITY

In the previous discussions it has been assumed that the velocity within a layer is constant, but there are cases where the velocity increases linearly with depth according to the equation $V = V_0 + kz$, where V_0 is the velocity at the layer surface, V the velocity at depth z and k a constant. Continuous velocity increases of importance are rarely encountered in shallow refraction seismics. Sometimes it can be observed that in upper weathered and fractured rock layers the velocities gradually increase. Such cases can be solved by replacing the curved velocity line by a number of straight velocity segments.

4

Interpretation methods

The idealized geological conditions and corresponding time–distance–depth relations in the examples given in Chapter 3 are seldom encountered in reality. The recorded raw travel time curves are generally distorted due to abrupt changes in topography and refractor configuration, variations in thickness and number of overburden layers and varying velocities in horizontal and vertical directions. The dry and loose near-surface layers constitute a special problem since their velocities generally are very low and, therefore, relatively small variations in thickness of the topmost layers influence the recorded arrival times to a higher degree than more prominent deep features where the velocity is higher.

The problem to be solved is to eliminate or at least minimize the influence of varying geological conditions along the profiles and to transform the irregular, recorded travel time data into ideal cases so that depth and velocity determinations can be made.

4.1   DEPTH DETERMINATIONS

For depth determinations at the impact points and at the receiving stations methods particularly suited for solving shallow interpretation problems are analysed below. To obtain true intercept times at the impact points the law of parallelism and the ABEM correction method are proposed while the ABC method and Hales' method can conveniently be used for depth calculations at the receiving end of the trajectory. A considerable part of the analysis below is devoted to errors inherent in the various techniques.

4.1.1   The law of parallelism

One of the basic concepts in refraction seismics is parallelism. It can be used to obtain true intercept times at impact points as well as at receiving stations.

Moreover, parallelism, together with reciprocity, is very important for the analysis of the velocity picture. The basic idea is as follows.

As mentioned previously in Section 3.1.1, an intercept time consisting of two different delay times, one at the energy source and another at the detector position, cannot give a unique solution but only an average depth. However, a true intercept time can be obtained if the refractor velocity segment from an impact point is extrapolated back to the point parallel to a covering refractor velocity curve from a distant impact point.

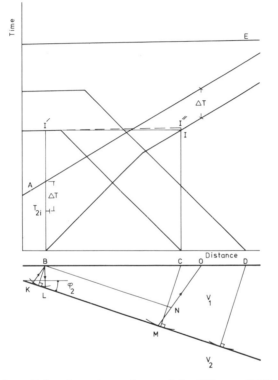

Fig. 4.1 Law of parallelism for a plane refractor (after Sjögren, 1980).

Curves A and B in Fig. 4.1 are used to demonstrate the procedure. The time difference ΔT between curve A and the refractor velocity segment of curve B is plotted at point B, as indicated in the figure. The intercept time T_{2i} thus obtained refers only to the conditions at B and the influence of the difference in depth between the impact point and the receiving station is eliminated. The arrival times from the offset impacts A (to the left) and E (to the right) are due to waves travelling in the V_2-layer. Note the parallelism between the refractor velocity curves in both directions and the reciprocity between the refractor velocity segments from B, C and D.

The graphical procedure can be explained in the following manner. The time difference between waves arriving at B from the distant impact A and those arriving at an arbitrarily chosen point O from the same impact is

$$KL/V_2 + LM/V_2 + MO/V_1 - KB/V_1 \qquad (4.1)$$

Since $KB = BL = MN$, the expression reduces to

$$KL/V_2 + LM/V_2 + NO/V_1 \qquad (4.2)$$

The travel time from B to O via the refractor is

$$BL/V_1 + LM/V_2 + MN/V_1 + NO/V_1 \qquad (4.3)$$

The mathematical significance of the projection of the refractor velocity segment in curve B back towards the impact point B parallel to curve A is that the travel time (4.2) is subtracted from the time (4.3). The result is the true intercept time T_{2i}.

Hence

$$T_{2i} = BL/V_1 + MN/V_1 - KL/V_2 \qquad (4.4)$$

where

$$BL = MN = \frac{h_1 \cos\varphi_2}{\cos i_{12}}$$

h_1 being the vertical depth at B. Also

$$KL = 2h_1 \cos\varphi_2 \tan i_{12}$$

and

$$V_2 = V_1 \sin i_{12}$$

Substituting, we obtain

$$T_{2i} = \frac{2h_1 \cos\varphi_2}{V_1 \cos i_{12}} - \frac{2h_1 \cos\varphi_2 \sin^2 i_{12}}{V_1 \cos i_{12}}$$

$$= \frac{2h_1 \cos\varphi_2 \cos i_{12}}{V_1} \qquad (4.5)$$

and

$$h_1 = \frac{T_{2i} V_1}{2\cos i_{12} \cos\varphi_2} \qquad (4.6)$$

If the factor $\cos\varphi_2$ is disregarded in Equation (4.6), we obtain the perpendicular distance to the refractor. A short arc of this radius is then plotted on the cross-section. The depths at C and D are calculated in the same manner

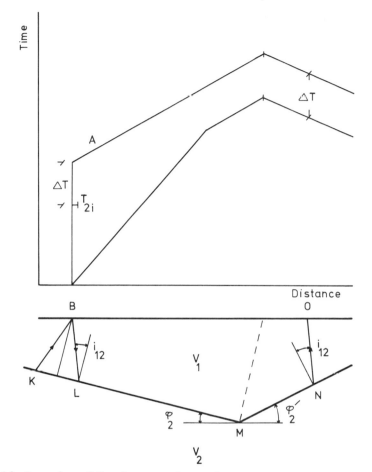

Fig. 4.2 Law of parallelism for a non-planar refractor.

using parallelism. A tangent to the respective arcs yields the refractor configuration.

The quantities in Equation (4.6) refer to the conditions close to the impact point, i.e. within the triangle BLK. The effect of factors outside this small area are eliminated. Therefore, sources of error, causing a distortion of the ideal triangle, are to be found in the immediate vicinity of the impact point. Errors will be introduced if, for instance, the refractor between K and L is non-planar or there is a velocity change within the triangle.

The example in Fig. 4.2 demonstrates that the law of parallelism is valid even if the refractor is non-planar between the impact point and the receiving end. The derivation of the equations is made in the same way as for Fig. 4.1.

The time difference of waves arriving at B and O from the impact point A is

$$KL/V_2 + LM/V_2 + MN/V_2 + NO/V_1 - KB/V_1 \qquad (4.7)$$

The total travel time from B to O via the V_2-layer is

$$BL/V_1 + LM/V_2 + MN/V_2 + NO/V_1 \qquad (4.8)$$

Thus

$$T_{2i} = (4.8) - (4.7) = BL/V_1 + KB/V_1 - KL/V_2 =$$

$$= \frac{2h_1 \cos \varphi_2 \cos i_{12}}{V_1} \qquad (4.9)$$

Solving for h_1, we obtain

$$h_1 = \frac{T_{2i} V_1}{2 \cos i_{12} \cos \varphi_2} \qquad (4.10)$$

Once again, the true intercept is obtained and it is dependent only on the conditions at the impact point. The dip change does not affect the intercept time obtained. In the same manner it can be shown that other factors outside the impact area such as, for instance, variations of elevation, varying velocities in the overburden or in the refractor, are automatically eliminated.

The concept of parallelism can also be used to give true intercept times at detector positions, as shown by D. W. Rockwell (1967, pp. 373–4). He makes use of a combination of reciprocity and parallelism. In Fig. 4.1 the intersection between the vertical time axis through point C and the refractor velocity segment from point B is designated I. According to the law of reciprocity, a horizontal line through I yields the reciprocal time above B, so that $I' = I$. If this time value I' is projected back to point C parallel to curve E, we obtain the intercept I'', which gives the intercept time T_{2i} for the impact point C. Point C was chosen for the above demonstration since there is a registration in the reversed direction, but any arrival time on the refractor curve from B can be utilized for a depth determination, provided that the overburden velocity is known.

For multilayer cases it is often not sufficient to establish only the intercepts of the bottom refractor. Velocities and intercepts for the overlying layers are likely to deviate from the true values. As long as there is a coverage in both directions of registrations of the various refractor velocities, the problem can be solved by an appropriate use of parallelism. Each velocity segment is projected back to the impact point parallel to an overlying curve from the same layer. The technique is described below.

The sloping interface between the first and second layers in Fig. 4.3 means that the recorded V_2 and V_3 velocities are apparent, being too low in the forward and too high in the reverse curves. It is assumed that the bottom refractor (V_3) is covered in both directions by registrations from the offset impact points A and C. There are also apparent velocities available from the second layer (V_2). These velocities, marked 3 and 4 in the graph, are assumed to have been obtained from nearby impacts in the same line. The travel time

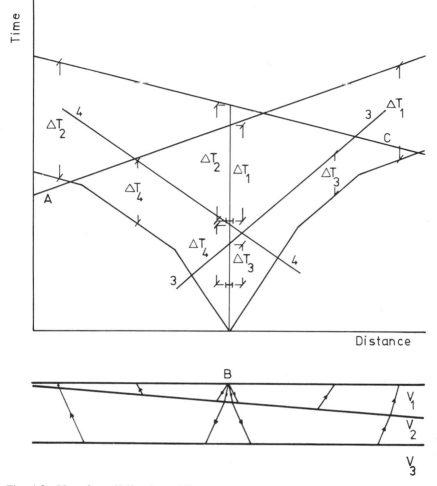

Fig. 4.3 Use of parallelism in multilayer cases.

curves from the second layer (3 and 4) and the third layer (A and C) are parallel to the respective velocity segments from point B. The true intercept time for the bottom refractor is obtained by projecting ΔT_1 and ΔT_2 to the point B, while curves 3 and 4 are used for the V_2-layer, time corrections ΔT_3 and ΔT_4.

It was mentioned previously that varying conditions in the vicinity of an impact point may act as sources of error. In Figs 4.4 and 4.5 the impact points lie close to a velocity change in the refractor and a neglect of this change leads to errors in depth. Errors caused by varying velocities are comparatively small but sometimes they augment other errors. In the figures it is assumed that $V_{2'}$ is less than V_2.

In Fig. 4.4 point B is the emergence of the wave front contact in the reverse recording, curve G. The direct recording gives rise to a diffraction zone between E and F due to the velocity boundary in the refractor. The recorded travel time curves from the impact points C and D refer to the ideal case; the interfaces are plane and horizontal, the slant raypaths from the impacts strike the refractor within the $V_{2'}$ region and the velocities V_1 and $V_{2'}$ are constant. Therefore, a direct prolongation of the refractor velocity segments from C and D back to the respective points gives the true intercept time $T_{2i} = 2h_1 \cos i_{12'}/V_1$, indicated by the lower short lines on the time axis through points C and D. On the other hand, if the parallelism between curve A and the refractor velocity segments from C or D is used to establish the intercepts, these deviate from the true ones. The erroneous intercept times are marked by the upper short lines on the time axis above the points C and D. If the latter

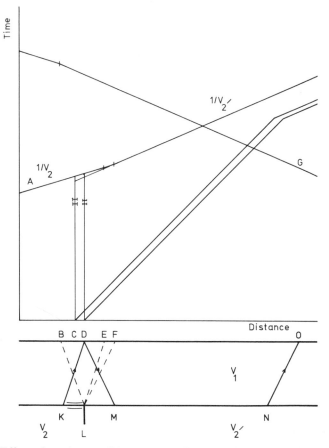

Fig. 4.4 Effect of varying conditions near the impact point, waves travelling from V_2 to $V_{2'}$ h_1 is the depth (after Sjögren, 1980).

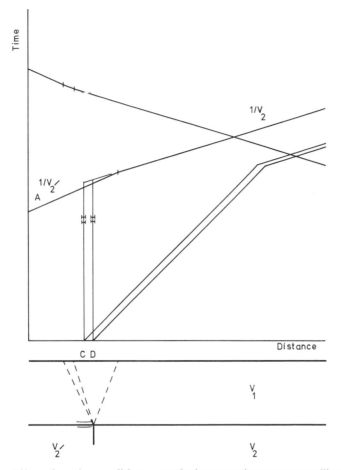

Fig. 4.5 Effect of varying conditions near the impact point, waves travelling from $V_{2'}$ to V_2 (after Sjögren, 1980).

intercept times are employed for the depth calculations neither V_2 nor $V_{2'}$ used in the depth equations yields the true depth. Using V_2 (or $V_{2'}$) under-estimates (or overestimates) the depth. The calculated depths are shown by arcs below point C.

The error in intercept time is eliminated if the $V_{2'}$ line segment in curve A is prolonged back to the impact points C and D. If this construction line is employed, the parallelism between the refractor curves results in true intercept times. These intercepts and the velocities V_1 and $V_{2'}$ give the true depths. The error, when the recorded curve A is used, is caused by the fact that the parallelism between the curves from A and from C and D ends at F, the right-hand boundary of the diffraction zone. The raypaths from C and D lie completely within the $V_{2'}$ area, while the slope of curve A is $1/V_2$ above C and D.

For an analysis of the source of error, we choose the curve from the impact point D in Fig. 4.4. The time differences between D and O for waves arriving from A is

$$KL/V_2 + LM/V_{2'} + MN/V_{2'} + NO/V_1 - KD/V_1 \qquad (4.11)$$

The travel time via the refractor from D to O is

$$DM/V_1 + MN/V_{2'} + NO/V_1 \qquad (4.12)$$

By subtracting (4.11) from (4.12), we get an apparent intercept time

$$T_{2iapp} = DM/V_1 + KD/V_1 - KL/V_2 + LM/V_{2'} \qquad (4.13)$$

where

$$DM = h_1/\cos i_{12'}$$

$$KD = h_1/\cos i_{12}$$

$$KL = h_1 \tan i_{12}$$

and

$$LM = h_1 \tan i_{12'}, h_1 \text{ being the depth.}$$

Thus, after simplification

$$T_{2iapp} = \frac{h_1 \cos i_{12}}{V_1} - \frac{h_1 \cos i_{12'}}{V_1} \qquad (4.14)$$

The intercept time obtained consists of two separate and different delay times. In order to get the true intercept time, $2h_1 \cos i_{12'}/V_1$, the right-hand side of Equation (4.14) has to be adjusted by $-(h_1 \cos i_{12}/V_1 - h_1 \cos i_{12'}/V_1)$, which is the time difference at D between the recorded travel time and the construction line. On both sides of D, the conditions are more complicated. The errors decrease to the right and increase to the left of D. The turning points are B and F where the true depths are obtained.

Figure 4.5 shows a structure similar to that in Fig. 4.4 but the wave propagation is from the lower to the higher refractor velocity. If the Law of Parallelism is applied to curve A and the refractor velocity curves from C and D, erroneous depths are obtained. The travel times for points C and D are due to the V_2 region while curve A above the points refers to $V_{2'}$. The intercepts when the recorded curve A has been used for correction are given by the lower short lines on the time axes. To get true intercepts, a construction similar to that in Fig. 4.4 can be applied to curve A. The slope $1/V_2$ is prolonged back to points C and D. A renewed interpretation, using this construction, yields the true intercepts. The velocities V_1 and V_2 have to be

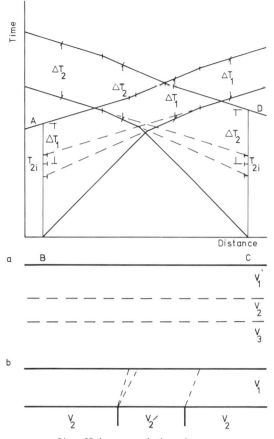

Fig. 4.6 Consequence of insufficient travel–time data.

used to get the correct critical angle i_{12} in the depth formula. The depths referring to the apparent intercept time are plotted below C. Using the velocities V_1 and V_2 leads to an underestimation of the depth giving the upper arc. A use of the velocities V_1 and $V_{2'}$ results in a slight overestimation, the arc below the refractor surface.

Errors in depth caused by varying refractor velocities depend on the interpretation approach, choice of refractor velocity and location of the impact point in relation to the velocity boundary. The errors are insignificant when the refractor velocities differ only slightly. More serious is the fact that the calculations tend to underestimate the depth above marked low-velocity zones in the refractor.

The concept of parallelism, besides providing true intercept times, is vital for the analysis of the subsurface geological conditions, for instance, for determining the number of velocity layers. With a sufficient amount of registrations, much misinterpretation can be avoided. If in Fig. 4.6 we assume that the curves A and D, the arrival times of which are due to the refractor, are missing, it is likely that the three velocity segments from the impact points B and C are interpreted as a three-layer case. This solution is indicated by the prolongation of the velocity segments (dashed lines) to the respective impact points. The corresponding depth and velocity analysis is given in (a). However, the refractor velocity curves A and D reveal that there is a velocity change in the refractor and that the second and third velocity segments from B and C are parallel to the respective overlying refractor velocity curve, as exemplified by ΔT_1 and ΔT_2 on the graph. The correct interpretation of the structure is to be found in (b). The true intercepts, T_{2i}, have been obtained by projecting in the usual manner the time differences ΔT_1 and ΔT_2 to the impact points. Because of insufficient information, namely the assumed missing curves A and D in this case, a two-layer case with a velocity change in the refractor was mistakenly interpreted as a three-layer case.

The curves from B and C in Fig. 4.6 can also correspond to other two-layer structures, for instance, a ridge in the refractor in the central part of the profile, in which case the second velocity segments are due to the down-dip sides of the ridge, or a depression in the refractor where the higher velocities in the third segments refer to up-dip parts of the depression. In the absence of sufficient information it is thus possible to obtain, in this particular case, at least four different solutions to the travel time curves from B and C.

4.1.2   The ABEM correction method

In the late 'forties the ABEM Company in Sweden introduced a special correction technique for depth determinations at the impact points in order to overcome interpretation problems caused by heavily distorted travel time curves due to rough terrain with greatly varying depths and velocities.

The interpretation procedure can be demonstrated most conveniently in a graphical manner. Through the intersection between the time axis at the impact point B in Fig. 4.7 and an overlying velocity curve A, referable to the refractor, a correction line (shown as a thin line) is drawn with a slope equal to the inverse of the refractor velocity V_2. Time differences, such as ΔT in the figure, between the recorded refractor curve A and the correction line are then used to correct the arrival times at the receiving points on the refractor velocity segment from the impact point B. The dots in the time–distance graph represent the corrected travel times. The line through these dots has the same slope ($1/V_2$) as the slope of the correction line and is then produced towards the impact point B where T_{2i} yields the true intercept time. It is assumed that in all the examples given later, the depth interpretation has been preceded by a detailed refractor velocity analysis according to Section 4.2.1. The velocity V_2 has to be used for the depth formulae, intercepts or critical distances, since the two delay times composing the intercept time refer to this velocity.

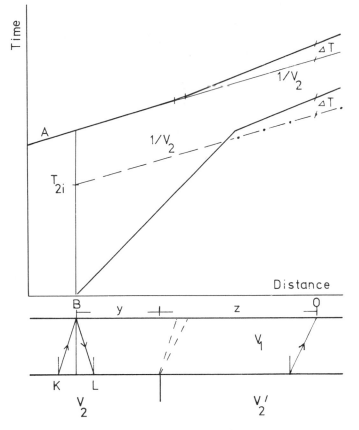

Fig. 4.7 Principle of the ABEM correction method.

In Fig. 4.7 the layers are horizontal and plane. The velocity in the upper layer is assumed to be constant but there is a velocity change in the refractor, with $V_{2'} < V_2$.

Since the angles of incidence and emergence are different, the time–distance–depth relation for arrivals from the impact point B is

$$T_2 = \frac{h_1 \cos i_{12}}{V_1} + \frac{h_1 \cos i_{12'}}{V_1} + \frac{x}{V_2} \qquad (4.15)$$

The apparent intercept time is

$$T_{2\text{iapp}} = \frac{h_1 \cos i_{12}}{V_1} + \frac{h_1 \cos i_{12'}}{V_1} \qquad (4.16)$$

The intercept is composed of two different delay times, one at the impact point and the other at the receiving end of the trajectory. In order to have an intercept time of two identical delay times, the influence of the different velocities in the refractor has to be eliminated.

The time difference between arrivals at B from the distant impact A and the arrivals at the arbitrarily chosen point O from the same impact is

$$\frac{h_1 \tan i_{12}}{V_2} + \frac{y}{V_2} + \frac{z}{V_{2'}} - \frac{h_1 \tan i_{12'}}{V_{2'}} + \frac{h_1}{V_1 \cos i_{12'}} - \frac{h_1}{V_1 \cos i_{12}} \tag{4.17}$$

while the time difference for the correction line with the constant refractor velocity V_2 over the same distance is

$$\frac{y}{V_2} + \frac{z}{V_2} \tag{4.18}$$

The total travel time from B to O via the refractor is

$$\frac{h_1}{V_1 \cos i_{12}} + \frac{y}{V_2} - \frac{h_1 \tan i_{12}}{V_2} + \frac{z}{V_{2'}} - \frac{h_1 \tan i_{12'}}{V_{2'}} + \frac{h_1}{V_1 \cos i_{12'}} \tag{4.19}$$

The correction ΔT in Fig. 4.7 is obtained by subtracting (4.18) from (4.17). Thus

$$\Delta T = \frac{h_1 \tan i_{12}}{V_2} - \frac{h_1 \tan i_{12'}}{V_{2'}} + \frac{h_1}{V_1 \cos i_{12'}} - \frac{h_1}{V_1 \cos i_{12}} + \frac{z}{V_{2'}} - \frac{z}{V_2}$$

The time in (4.19) minus the correction term ΔT gives the following equation

$$T_2 = \frac{2h_1}{V_1 \cos i_{12}} - \frac{2h_1 \tan i_{12}}{V_2} + \frac{y}{V_2} + \frac{z}{V_2}$$

After simplification and replacing $y + z$ by x, we obtain

$$T_2 = \frac{2h_1 \cos i_{12}}{V_1} + \frac{x}{V_2} \tag{4.20}$$

which is the ordinary equation for a two-layer case with horizontal interfaces and constant velocities in overburden and refractor. The correction method has thus removed the influence of the velocity change in the refractor and

$$h_1 = \frac{T_{2i} V_1}{2 \cos i_{12}} \tag{4.21}$$

Other distortions of the recorded travel times are also eliminated by this correction technique. In Fig. 4.8 the method is demonstrated for a case with an irregular refractor surface. The corrections are made in the same manner as described above. A correction line of slope $1/V_2$ is drawn from a point of the refractor velocity curve A vertically above point B. The slope angles of the

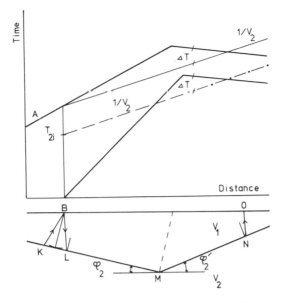

Fig. 4.8 ABEM correction method applied to an irregular refractor surface.

refractor need not be equal but the velocity in the refractor is assumed to be constant and the waves are critically refracted into the V_1-layer.

The time difference between arrivals at B and O of waves from the offset impact A is

$$KL/V_2 + LM/V_2 + MN/V_2 + NO/V_1 - KB/V_1 \tag{4.22}$$

while the corresponding time difference for the correction line is

$$BO/V_2 \tag{4.23}$$

The travel time from B to O via points L, M and N on the refractor is

$$BL/V_1 + LM/V_2 + MN/V_2 + NO/V_1 \tag{4.24}$$

According to the graphical construction, $(4.24) - [(4.22) - (4.23)]$ yields the corrected travel time T_2.

Thus

$$T_2 = BL/V_1 + KB/V_1 - KL/V_2 + BO/V_2$$

where

$$BL = KB = 2h_1 \cos\varphi_2/\cos i_{12}$$

$$KL = 2h_1 \cos\varphi_2 \tan i_{12}$$

$$BO = x$$

and h_1 is the vertical depth below point B. Finally

$$T_2 = \frac{2h_1\cos\varphi_2\cos i_{12}}{V_1} + \frac{x}{V_2} \qquad (4.25)$$

When $x = 0$, $T_2 = T_{2i}$ and

$$h_1 = \frac{T_{2i}V_1}{2\cos i_{12}\cos\varphi_2} \qquad (4.26)$$

If the factor $\cos\varphi_2$ is disregarded, the formula gives the perpendicular distance to the refractor which can be plotted as an arc on the cross-section below B. A tangent curve to the arcs at the impact points with their respective calculated depths yields the refractor configuration.

The correction technique demonstrated gives the true intercept times. Any influence of varying geological conditions such as, for instance, variations in velocities, depths, elevation and number of layers away from the immediate vicinity of the depth determination points is automatically eliminated since the corresponding time terms are included in the travel times from the impact points as well as in the correction quantities but with different signs. One objection to the correction method may be that the correction line is drawn without considering the inclination of the refractor but this source of error is eliminated since the corrected refractor velocity segment is produced towards the depth determination point with the same slope $1/V_2$ as for the correction line. The law of parallelism and the ABEM correction method yield identical intercept times, namely the true ones. In both cases possible sources of error are to be found within the triangle formed by the impact point and the two foot points on the refractor for the critical rays, KBL in the figures. Velocity variations and a non-planar refractor within the 'triangle of error' may affect the accuracy of the depth determinations. These questions will be dealt with in Section 4.1.3.

In the examples given above, the true intercept times of the bottom refractor have been obtained. However, the ABEM correction technique can also be used to adjust irregular travel times from overlying layers with the aid of a velocity curve from an underlying layer. The procedure to correct recorded travel times by a curve from an underlying layer must be used with great caution in order not to introduce erroneous velocities or even non-existent velocity layers. In fact, it is mathematically correct to adjust a velocity curve only by arrival time data from the same layer. Nevertheless, the technique greatly facilitates the analysis and interpretation of badly distorted travel time curves and sometimes it is the only way to solve an otherwise unsolvable interpretation problem.

Figure 4.9 shows a case with varying elevation while the interfaces between the velocity layers are assumed to be plane and horizontal. In this case the recorded arrivals from the third as well as the arrivals from the second layer

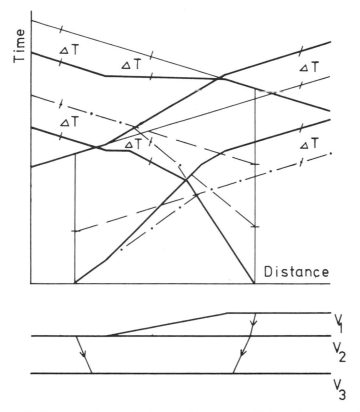

Fig. 4.9   ABEM correction method applied for a case with irregular topography.

have been corrected by means of the apparent V_3 curves. The slope of the correction lines employed is $1/V_3$. In the region with the inclined ground surface, the recorded travel time curves yield velocities which deviate considerably from the ideal ones. In this case the errors in velocity for the second layer V_2 are 25 and 40% in the forward and reverse recording respectively. The model in the figure is constructed with velocities V_1, V_2, and V_3 in the proportions 1, 2, and 5 respectively.

Because of the different slopes of the slant raypaths from the V_2- and V_3-layers, there is a residual error in the calculated velocity for V_2 even after correction. The technique of using correction terms based on the apparent V_3 velocities leads in this particular case to an overestimation of V_2 in the direct and an underestimation in the reverse recording. In the usual manner the dots designate the corrected arrival times and the correction time terms are given by ΔT. The residual error in V_2 is about 5%. However, in practical work the evaluation of the velocities is based not only on corrected arrival times but also is a synthesis of recorded and corrected velocities, averages of the velocity values and information from other impact points and

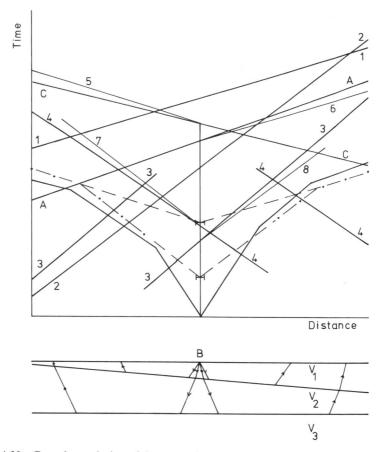

Fig. 4.10 Complete solution of the correction problem.

profiles. The interpretation problem regarded from a purely construction viewpoint is to establish the slopes of the velocity lines and the location of these lines on the time–distance graph to satisfy the local geological conditions at a particular impact point. Generally, a close approximation to the reality, but not a perfect match, can be achieved.

A mathematically exact solution of the correction problem is presented in Fig. 4.10. In this example there is complete coverage in both directions of apparent velocities from the second and third layers. The travel time curves A and C correspond to the V_3-layer. Curves 3 and 4, the V_2-layer, are assumed to have been obtained by connecting apparent V_2-layer velocity segments from point B and adjacent impact points.

The inclination of the interface between the first layer and the second causes a deviation in the recorded velocities from the second and third media from the true values, higher in the reverse and lower in the direct recording.

Curves 1 and 2 refer to the harmonic mean velocity determinations (see Section 4.2.1) of V_3 and V_2 respectively.

To obtain a true intercept time for the V_3-layer, correction lines 5 and 6 have been drawn from the intersections between curves A and C and the time axis through point B. The slope of lines 5 and 6 is equal to that of curve 1, namely the estimate of $1/V_3$. In the same manner the time differences between the correction lines 7 and 8, with the slope of curve 2, and the apparent V_2 velocity curves 3 and 4 have been used to correct the V_2 segments from the impact point B. Unfortunately, complete velocity coverage of all the layers composing the subsurface is encountered rather seldom. In such a case second arrivals, if detectable in the records, can provide missing travel time curves.

In Figs 4.11 and 4.12 there is a velocity change in the refractor, with $V_{3'} < V_3$, and the interface between the V_1- and V_2-layers is inclined in

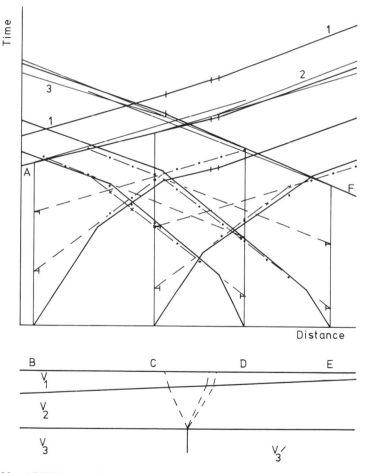

Fig. 4.11   ABEM correction method for irregular conditions.

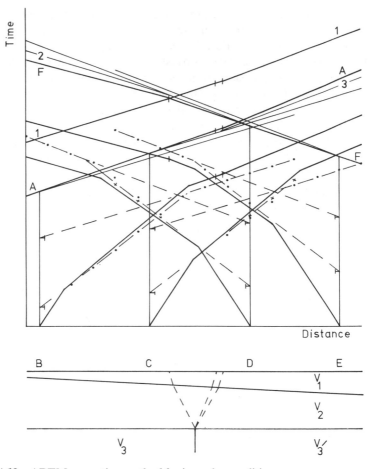

Fig. 4.12   ABEM correction method for irregular conditions.

different directions. The velocity boundary in the bottom refractor causes a diffraction zone in the direct recording and a wave front contact in the reverse direction. The zone and contact are given by dashed lines in the cross-section and are indicated by short lines in the refractor velocity curves A and F. Curve 1 in the graphs indicates the computed mean velocity of the third layer. The velocity determination procedure will be explained in Section 4.2.1.

There is no coverage of apparent V_2 velocities so that the corrections of the recorded velocity curves from the second and third layers must be based on apparent V_3 and $V_{3'}$ velocities. At points B and C in Fig. 4.11, the correction lines are given the slope $1/V_3$, where V_3 is obtained from curve 1, since these impact points are located within the V_3 area and the rays from the points strike the refractor to the left of the boundary separating the V_3 and $V_{3'}$ sections. For the same reason the corrections at points D and E have been based on the velocity $V_{3'}$. True intercept times and velocities for the bottom refractor are obtained. The dots on the graph show the corrected arrival times. As in the example in Fig. 4.9 there is a residual, but insignificant, error in the determination of V_2 at points B and E. The serious errors in the determination of V_2 are found in the travel time curves from C and D. The corrections have been made without regard to the velocity change in the refractor. For point C, for instance, the velocity line segment for the second layer lies within the $V_{3'}$ area while the correction line has a slope of $1/V_3$. Because of the slope of the interface between the upper layer and the second, the apparent velocity for the V_2-layer recorded from C is higher than the true velocity. If the corrected arrival times, the dots, are used to construct the velocity line, the error increases even more. The proper correction technique in this case is to change the inclination of the correction lines, in accordance with the velocity of the appropriate section of the third layer, at the emergence of the diffraction zone and the wave front contact. The modified correction procedure is given in the graph by lines 2 and 3. A renewed correction is marked by crosses.

The same reasoning as for Fig. 4.11 is valid for the example presented in Fig. 4.12 in which the interface between the V_1- and V_2-layers is dipping to the right. Other pertinent data are the same as for Fig. 4.11. When correcting the apparent V_2 velocity segments, the correction lines must be adjusted to the slope of the actual refractor velocity. The correction line used for the registrations from point C changes the slope from $1/V_3$ to $1/V_{3'}$ at the diffraction zone, line 3. In the other direction (point D), the slope for the correction line is $1/V_3$ to the left of the wave front contact, line 2. The crosses indicate the corrected travel times after the velocity change in the refractor has been considered.

It can be seen that the time corrections are considerably greater in Fig. 4.12 than in Fig. 4.11. In the latter figure the difference in the velocity in the third layer is partly counterbalanced by the decrease in thickness of the V_1-layer while in Fig. 4.12 there is a cumulative effect of the increase in thickness of the V_1-layer and the lower velocity $V_{3'}$.

Figure 4.13 demonstrates the importance of covering the entire traverse to be investigated by refractor velocities from both directions. If the refractor velocity curves A and D were missing, the straightforward interpretation of the travel time curves from B and C would lead to a three-layer case. However, this interpretation problem is solved by curves A and D. The corrected arrival times from B and C lie on straight lines with the same slope $1/V_2$ as that used for the correction lines. Another feature is that the refractor velocity curves A and D are parallel to the corresponding velocity segments from B and C. Ultimately, the curve of the mean refractor velocity (1) reveals that in the central part of the profile there is a section with a velocity lower than V_2. Curve 1 will be referred to in connection with the mean refractor analysis in Section 4.2.1. Since in this case the low-velocity zone lies outside

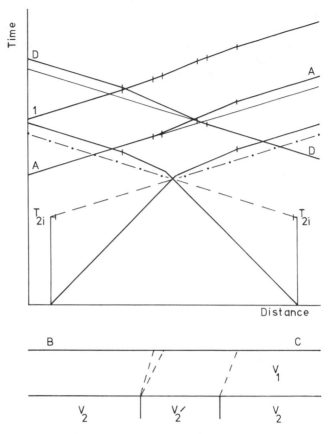

Fig. 4.13   Importance of complete refractor velocity coverage.

the critical regions at B and C, true intercept times at B and C can be obtained.

The ABEM method of corrections for a variety of structures is shown qualitatively in Figs 4.14–4.19. Time deviations are simplified and the effect of different raypaths for the waves from the V_2- and V_3-layers is not considered. It is assumed in all cases that the apparent velocity curves from the second and third layers have to be corrected by the aid of the arrival times from the third layer, i.e. from V_3 curves. The main objective is to study the corrections of the intermediate layer. It is seldom necessary to correct the velocity of the top layer.

Figure 4.14

Because of the elevation of the ground, the recorded travel time curves give a false picture of the geology. The velocity lines, magnitude of velocity and position on the time–distance graph do not represent the subsurface conditions at impact points B and C. The varying ground surface causes an underestimate of the velocity V_2 from point B and an overestimate from C.

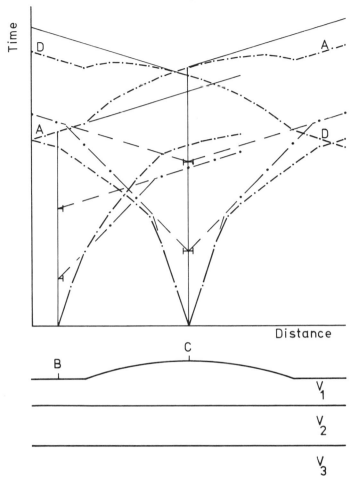

Fig. 4.14 ABEM correction method for a case with ground elevation.

The arrival times from the third layer refer, not to the conditions at the impact points, but to an average along the traverse.

The correction lines, slope $1/V_3$, are applied to the V_3-layer curves A and D in the usual manner and the curves are corrected as shown by dots. It is likely that the corrected arrival times still do not give the true velocity V_2 of the second layer or not even the same velocity in different directions. For a final decision the interpreter has to evaluate the recorded curves, the corrected ones, average velocities and the intercepts in different directions.

Figure 4.15

The depression in the ground surface causes an overestimate of the velocity of the second layer from B and an underestimate from C. In principle, the same reasoning is valid in this case as for Fig. 4.14.

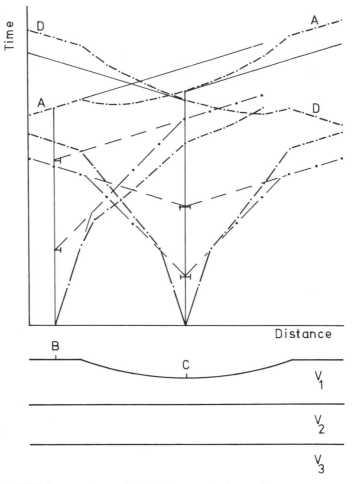

Fig. 4.15   ABEM correction method in the case of a depression.

Figures 4.16 and 4.17

In these examples the interface between the first layer and the second is assumed to be non-planar. The varying interface affects the registrations from the second and third layers. Note that, contrary to the cases given in Figs 4.14 and 4.15, the dip changes in the V_2-layer and the recorded velocity anomalies display different horizontal positions because of the slant raypaths. Moreover, the places for the velocity changes in curves A and D are displaced in relation to each other. In these simplified examples the corrected points lie on straight lines. In reality, because of varying layer thicknesses and different raypaths from the second and third layers, the corrected arrival times lie slightly below or above the true values.

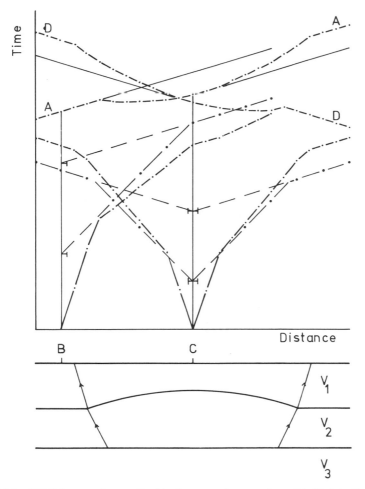

Fig. 4.16   ABEM correction method in the case of a non-planar V_1–V_2 interface.

Figures 4.18 and 4.19

The travel times from the third layer V_3 have to be corrected. The recorded arrival times from the second layer form straight lines and, moreover, the velocities are equal in both directions. Judging from the simplified velocity picture, it is obvious that there is no need to correct velocity segments other than those from the third layer and serious errors will be introduced if the arrival times from the second layer are corrected. In reality, however, ideal curves are encountered rather seldom and the influence on the recorded arrivals of irregularities in the bottom refractor may be obscured by varying conditions in the upper layers. Perfunctory use of the correction methods described may lead to erroneous interpretations, depths, velocities, number

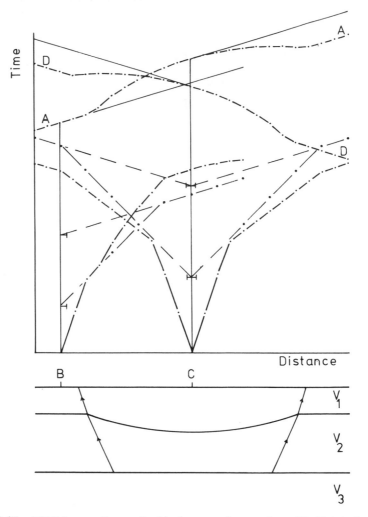

Fig. 4.17   ABEM correction method in the case of a non-planar V_1–V_2 interface.

of velocity layers, etc. After the first rough interpretation and the drawing of the corresponding structure, it is expedient to re-evaluate the analysis. Does the structure obtained fit the recorded travel times and the total velocity picture? To what extent have geological conditions in a layer, the velocity of which has been used as a basis for the corrections, affected the adjusted arrival times in the overlying layers?

An application of the correction methods is presented in Fig. 4.20. The total length of the profile is 105 m. The geophones were placed in a regular system with a separation of 5 m. A 12-channel set of equipment was used for the measurements. In order to cover the profile, the survey had to be carried

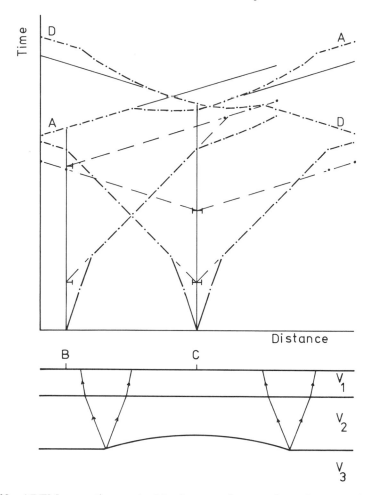

Fig. 4.18 ABEM correction method in the case of a non-planar bottom refractor surface.

out in two steps since one 12-channel geophone spread with 5 m spacing gives 55 m. In the first layout, 0–55 m, all shots were fired, including the shots A and B beyond the ends of the profile. The last two geophones in the first layout were used for the overlapping of the travel times, i.e. the first geophone of the second spread started at 50 m in the profile. The end-point will then be at 105 m. In the second spread all shots were repeated and the arrival times at the overlapping geophones were used to tie in the travel time curves. The various components of the continuous profiling are present here, namely the shots inside the geophone layout, the offset shots A and B, the repeat recording and the uniting of separate layouts into one profile.

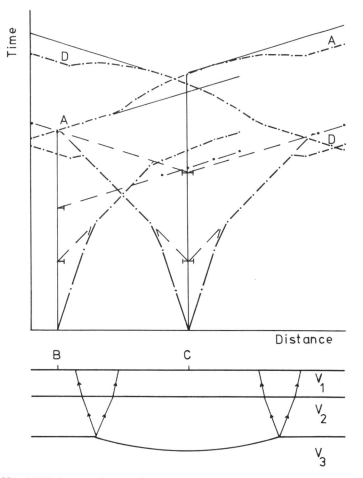

Fig. 4.19 ABEM correction method in the case of a non-planar bottom refractor surface.

The first step of the interpretation, when the time plotting is completed, is to check the overall time–distance picture, the parallelism between the curves, the reciprocity of the curves, an assumption of the number of layers, outstanding features in the velocities, etc. For the case in question we can summarize as follows: a simple two-layer case (only two types of velocity), the velocity of the overburden (1700 m/s) is equal in both directions, the slope of curve A in the beginning of the profile is negative while curve B is approaching the velocity of the overburden. The latter velocity features reveal that from about 50 m in the profile the bedrock is steeply sloping to the left.

The mean bedrock velocity – obtained by an averaging of the time increments – is indicated by curve 1. In the beginning of the profile the

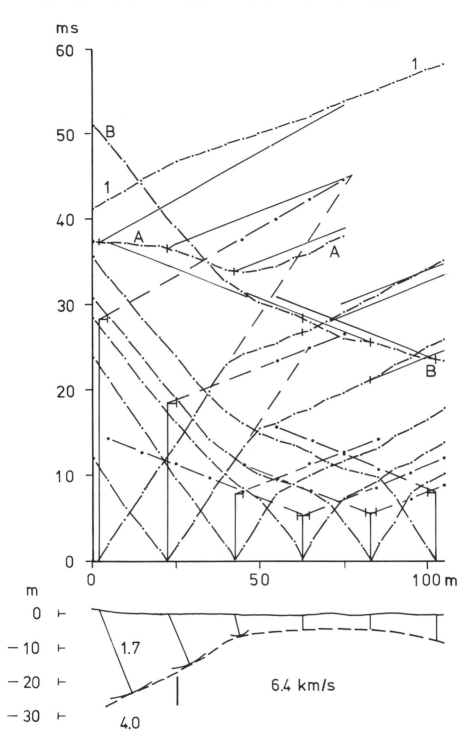

Fig. 4.20 ABEM correction method applied to a field example (Archives of ABEM Company, Sweden).

computed velocity is about 4500 m/s. The higher velocity, 6400 m/s, from 25 m onwards, corresponds to a diabase dyke. For the corrections, the lower velocity (4500 m/s) has been used for the first shotpoint in the beginning of the profile. The bedrock curves from the other shotpoints have been corrected by lines with the reciprocal slope of 6500 m/s. Note that in the figure the velocities are given in kilometres per second and the times in milliseconds.

The corrections for the first two shotpoints are considerable while from the profile length 50 m they are moderate, except for the reverse curve from the shotpoint to the right of 50 m. If the raw travel times from the first shotpoint had been used without corrections for a calculation, the depth would have been underestimated by about 35%. This smaller depth refers to a mid-point between the shotpoint and the apparent break point. For a depth calculation either the intercept times or the critical distances can be used according to the formulae in Sections 3.1.1 or 3.1.2 for a two-layer case. The bedrock surface is plotted in the cross-section as an envelope to the arcs at the shotpoints. The example of Fig. 4.20 will be further analysed in connection with the ABC method in Section 4.1.4.

The configuration of the bedrock in Fig. 4.20 is the cause of some errors. The rays in the direct recording make an angle with the bedrock surface to the right of the profile length 50 m and, consequently, in this area the waves are non-critically refracted to the ground surface. On the other hand, the rays from the reverse recording make an angle with the dipping part of the bedrock to the left of point 50 m. Therefore, the waves are non-critically refracted through the overburden. As a result of these sources of error, it can be anticipated that:

(a) The depth at the shotpoint to the left of 50 m is overestimated and
(b) The bedrock velocity values are slightly overestimated.

These questions are to be discussed in more detail in Section 4.1.3. The calculated mean velocity (4500 m/s) at the beginning of the profile is probably too high. In the cross-section 4000 m/s is given as an estimate of the actual velocity.

4.1.3 Some sources of error

As mentioned in Sections 4.1.1 and 4.1.2, a velocity or dip change in the refractor in the area where the slant raypaths leave or strike the refractor is critical and a neglect of these circumstances can lead to errors in the estimation of the intercept times and thereby also of the depths. In this section I have used the ABEM correction method to analyse the source and nature of some errors of depth at the impact points and to suggest some remedies to increase the accuracy of the depth determinations. The ABEM method is applied here to the problems, but the law of parallelism used for corrections will give the same errors.

The errors in depth caused by varying refractor velocities are generally moderate. They are, however, more pronounced above low-velocity zones. A seismic survey tends to underestimate the depths. Such a case is demonstrated in Fig. 4.21. The model is based on the relative velocities of 1, 2.33, and 3.33 for V_1, $V_{2'}$, and V_2 respectively. The travel times between the impact point B and the detector position O are considered. In spite of the fact that point B is situated above the $V_{2'}$ region, the travel paths from B to O have no connection with the lower velocity. The path in the refractor lies within the V_2-layer and since the depth h_1 is constant, the raypaths BE and FO in the overburden are

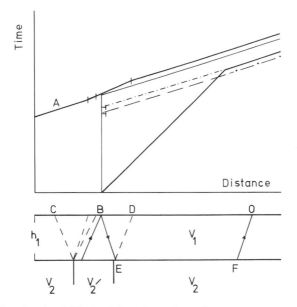

Fig. 4.21   Situation in which depth is underestimated.

equal, the critical angle being i_{12}. Therefore, the delay times at both ends of the trajectory are identical and a prolongation of the refractor velocity segment back to B would yield the true intercept time (the dash-dotted line). If the low-velocity zone is overlooked in the interpretation and a correction line with the slope $1/V_2$ is applied to curve A and corrections are made in the usual way, the depth will be underestimated. In this particular case, the computed depth is about 5% smaller than the true depth. To get the true intercept time or critical distance, the slope of the correction line has to be $1/V_{2'}$ between B and D, the emergence point of the wave front contact, and $1/V_2$ to the right of D. This correction line coincides with the recorded curve, and the correction terms will be zero in accordance with the assumption that curve B is ideal and therefore in no need of any corrections. The error when correcting is caused by the fact that the parallelism between curves A and B is lost at point D.

To the right of B in Fig. 4.21, the error decreases and at D the true depth is obtained. To the left of point B, the situation is a little more complicated. If the zone is wide enough in relation to the depth to the refractor, there is a region where the rays from the impact point B strike within the $V_{2'}$ section and curve A refers to $V_{2'}$ above point B. The true depth is then obtained if the reciprocal slope of $V_{2'}$ is used for the correction line and the depth formula. Further to the left, as can be seen in the figure, the rays from the impact points are within the $V_{2'}$ section while the overlying curve A refers to the higher velocity V_2. If V_2 is used for corrections and calculations, the depths are smaller than the true depths. An improvement in the calculations can be achieved by using $V_{2'}$ instead of V_2. At C, which is the wave front contact on the ground in the reverse recording, the true depth will be obtained if V_2 is used for corrections and depth calculations.

Numerous drilling results have shown that seismically determined depths of low-velocity zones are too shallow. Sometimes the deviations amount to 10–15% of the actual depth, particularly if the zones are narrow in relation to depth and the velocities are extremely low. The errors are only partly caused by the interpretation technique chosen. A dominant factor is that the fractured and weathered rock material in a shear zone is more eroded than the surrounding compact rocks, resulting in a local depression in an otherwise plane or moderately undulating bedrock surface. Some interpreters increase the depths calculated above low-velocity zones in accordance with experience of the local geological conditions. If the low-velocity zone is associated with a prominent depression in the bedrock, the discrepancies between calculated and actual depths can be considerably greater than that stated above. However, when comparing results from drilling and seismics, we must always bear in mind the difficulty, when drilling, in establishing accurately the boundary between boulders in the bottom soil layers and the upper parts of a highly fractured rock material in a shear zone.

When there are marked dip changes in the refractor, it is very likely that the seismically determined depths deviate from the actual depths. Critical parts can be found in association with, for instance, large scale faulting, ridges or depressions in the refractor.

In Fig. 4.22, depicting a symmetrical depression in the refractor, a conventional interpretation approach to the problem will grossly underestimate the depths in the central part of the depression.

Correction lines, marked B and C, are applied in the usual manner to the refractor velocity curve A above the impact points B and C. The slope of the lines is $1/V_2$. The corrected arrival times give the intercept times T_{2i}. The corresponding depths are plotted vertically below the impact points, as shown by the upper arcs. The error in depth at point B is about 20% of the perpendicular distance BP to the refractor. The error in the vertical depth is

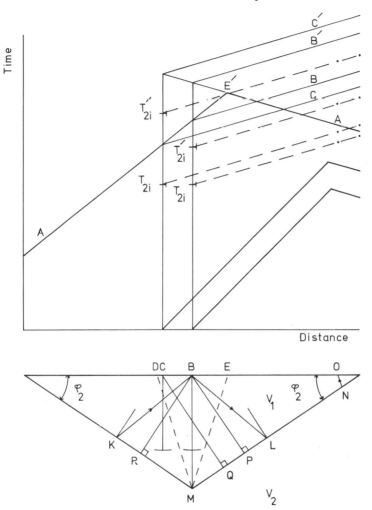

Fig. 4.22   Underestimation of depths due to depression.

about 34%. The intercepts and depths are the same at B and C. In fact, between D and E the same depth is obtained, namely the distance between these points and the refractor, perpendicular to the refractor. It can be mentioned that, for an evaluation of the errors, the model is based on the relative velocity values 1 and 3.33 for V_1 and V_2 respectively. The angle φ_2 is about 34°. The law of parallelism applied to the problem yields the same intercept time T_{2i}, or, more correctly expressed, T_{2iapp}, since it does not refer to the true depth.

The factors causing the error in depth can be analysed in the same way as previously used to prove the correction technique.

The time difference on curve A from B to O is

$$KM/V_2 + ML/V_2 + LN/V_2 + NO/V_1 - KB/V_1 \qquad (4.27)$$

The corresponding time increment on the correction line is

$$BO/V_2 \qquad (4.28)$$

The time from B to O along raypaths BL, LN and NO is

$$BL/V_1 + LN/V_2 + NO/V_1 \qquad (4.29)$$

The correction of the arrival times, taking the difference $(4.29) - [(4.27) - (4.28)]$, gives a time, designated $T_{2\text{app}}$, that does not refer to the actual depth below point B.

Thus

$$T_{2\text{app}} = BL/V_1 + KB/V_1 + BO/V_2 - KM/V_2 - ML/V_2$$

where

$$BO = x$$

$$MP = MR = h_1 \sin\varphi_2 \ (h_1 = BM)$$

$$BP = BR = h_1 \cos\varphi_2$$

$$PL = KR = h_1 \cos\varphi_2 \tan i_{12}$$

$$KM = ML = h_1(\sin\varphi_2 + \cos\varphi_2 \tan i_{12})$$

and

$$BL = KB = h_1 \cos\varphi_2/\cos i_{12}$$

so that finally

$$T_{2\text{app}} = \frac{2h_1 \cos\varphi_2}{V_1 \cos i_{12}} + \frac{x}{V_2} - \frac{2h_1 \sin\varphi_2}{V_2} - \frac{2h_1 \cos\varphi_2 \sin^2 i_{12}}{V_1 \cos i_{12}}$$

After simplification, we obtain

$$T_{2\text{app}} = \frac{2h_1 \cos\varphi_2 \cos i_{12}}{V_1} + \frac{x}{V_2} - \frac{2h_1 \sin\varphi_2}{V_2} \qquad (4.30)$$

and, when $x = 0$

$$T_{2\text{iapp}} = \frac{2h_1 \cos\varphi_2 \cos i_{12}}{V_1} - \frac{2h_1 \sin\varphi_2}{V_2} \qquad (4.31)$$

The error in the determination of the intercept time at B is caused by the second term on the right-hand side of the equation. This term corresponds to the travel time for the distance RMP. The travel paths from the impact point

B refer to the right section of the depression, while above point B the velocity segment on curve A emanates from the down-dip of the refractor to the left of B. The parallelism between curves A and B ceases at E', corresponding to the wave front contact E on the ground. In order to restore the parallelism, the velocity line segment in A corresponding to MN is extended back to the impact points B and C, a construction equivalent to second arrivals from MP or to an imaginary prolongation of MP to the left of M. Correction lines, B' and C', applied to the extension of curve A yield the true intercept times, marked T'_{2i} on the time plot. According to the law of parallelism, the time differences between curve A and the refractor velocity segments on curves B and C, plotted on the time axis through B and C from the construction line, give the same intercepts T'_{2i}. It should be noted that the calculated depths using the intercepts T'_{2i} have to be plotted normal to the refractor section ML, for B and C the points P and Q.

The depth analysis in the deeper part of the depression can be improved by an additional seismic line at B, perpendicular to the first profile. Such a profile will give the perpendicular distance to the refractor, BP or BR. The error in vertical depth at B is then reduced to about 17%, but still the area bounded by RMP is inaccessible for depth evaluations.

A low-velocity zone in the central part of the structure of Fig. 4.22 increases the error in depth. Such a case is shown in Fig. 4.23. Except for the velocity variation in the refractor in Fig. 4.23, other pertinent data are the same in the two figures. Because of the time delay in the low-velocity zone, the time term in Equation (4.31) causing the lowering of the calculated intercept times is increased. Consequently, the depths calculated in Fig. 4.23 are smaller than those obtained in Fig. 4.22. The underestimation of the distance from B to P is here about 26%. The corresponding figure is 20% in Fig. 4.22. Related to the depths obtainable is the distance between wave front contacts in the over-burden. This distance is greater in Fig. 4.23 than in Fig. 4.22 because of the cumulative effect of the depression and the low velocity in the refractor. The shadow zone in the structure of Fig. 4.23 and, therefore, also the error are bigger.

The true intercept time at B in Fig. 4.23 is obtained in the same way as described above. The prolongation back to point B of curve A is used for the correction line B'. The true intercept time, conforming to the perpendicular distance to the refractor, is designated T'_{2i}. The law of parallelism applied to the construction line gives the same intercept T'_{2i}.

The errors in a refractor depression increase with increasing slope of the sides and with decreasing velocity contrast between overburden and refractor. The combination of dip and velocity changes almost always represents serious problems with interpretation. The structure based on the first rough calculation is used for re-interpretation of depths and velocity distribution in the refractor to come closer to reality. These questions will be further treated in Section 4.1.6.

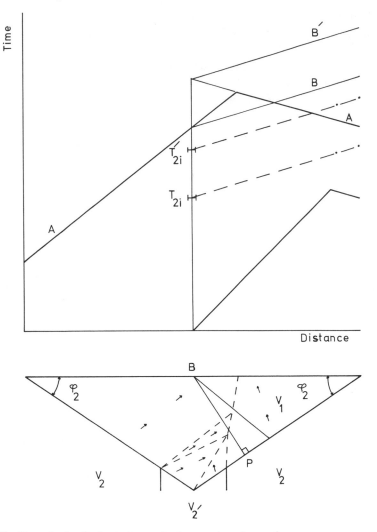

Fig. 4.23 Error in depth due to low-velocity zone in a depression.

There is a tendency for use of the first arrivals to cause the characteristic features of the refractor to be smoothed out. The smoothing-out effect is generally encountered in connection with abrupt changes of the refractor configuration and refractor velocity variations. Even rather moderate variations of the refractor surface may introduce sources of error such as, for instance, a loss of the parallelism between the curves involved in the depth determination, shortest travel paths through the structure and not along the surface resulting in non-critical refractions through the upper media, diffraction zones and curves with non-parallel raypaths. To a certain extent it

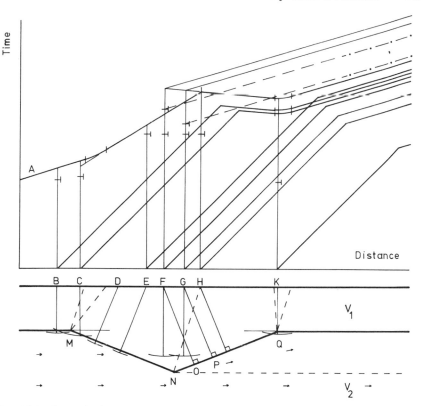

Fig. 4.24 Errors in depth, dip angle φ_2.

is possible to minimize the errors or at least estimate their magnitude and sign. The latter is a vital matter for evaluating the reliability of a seismic survey. The sources of error mentioned above are illustrated by the structures of Figs 4.24–4.27.

In Fig. 4.24 the computed depths are underestimated around N and over-estimated at the slope changes at M and Q. The errors at B and C are caused by a longer travel path in the refractor compared with the direction of the waves from the offset impact point A. Besides, at C there is a lack of parallelism between the curve from C and curve A. The raypaths from C refer to the inclined surface MN, while above C, curve A is composed of a velocity line from the horizontal part of the refractor and a diffraction zone to the right of point C. The error decreases to the right of C: see, for example, points D and E. The error at C can be diminished if the velocity line segment in curve A corresponding to MN is extended back to C as shown in the figure. An additional source of error is that the rays from A make an angle with the inclined refractor surface between M and N, causing non-critical refraction into the V_1-layer.

In the vicinity of the slope change at Q the calculated depths tend to be greater than the true depths. The waves from the depth determination points travel along the refractor surface and are critically refracted into the upper layer, while the waves from A impinge obliquely on the horizontal interface to the right of Q. These latter waves are non-critically refracted into the upper medium. There is an additional effect of the non-critical refraction, namely that a calculated mean refractor velocity will be somewhat higher than the true velocity V_2. The interpretation problem at point Q was encountered earlier, in Fig. 4.20. If the angle φ_2 is large, the recorded refractor velocity from K may be noticeably lower than the velocities from the impact points to the left of N, which can lead to the misinterpretation that the lower velocity is obtained in an upper layer underlain by a high velocity layer. The greatest errors are, however, to be found in the depression. The parallelism between curve A and curves F and G ends at H, the wave front contact on the ground surface. The shortest travel times from points F and G are along NQ, while the arrival times on curve A above F and G refer to the refractor to the left of N. This source of error can be eliminated if the velocity segment from NQ in curve A is extended to the depth determination points. Correction lines applied to the construction line give the upper intercepts, indicated by short lines on the time axes through points F and G. The corresponding depths have

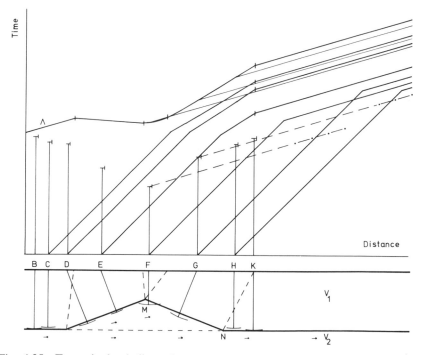

Fig. 4.25   Errors in depth dip angle φ_2.

to be plotted normally to NQ, the feet being O and P. An application of the law of parallelism gives the same results, errors as well as corrected depth values.

Also in the case of a ridge in the refractor, the features may be smoothed out. An interpretation making use of the first arrivals in a conventional manner overestimates the depth at the top of the ridge and underestimates the depths where the ridge comes to an end. Such a case is shown in Fig. 4.25. The errors at F and G are caused by:

(a) The different raypaths for the waves from A and those from F and G, and
(b) The rays from A making an angle with the refractor between M and N.

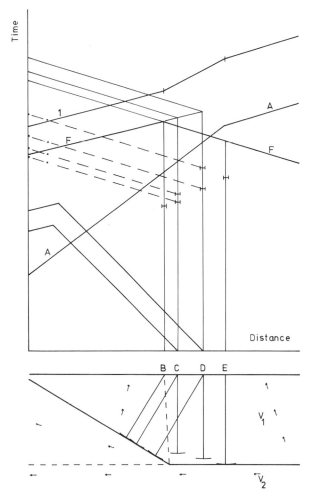

Fig. 4.26 Errors in depth.

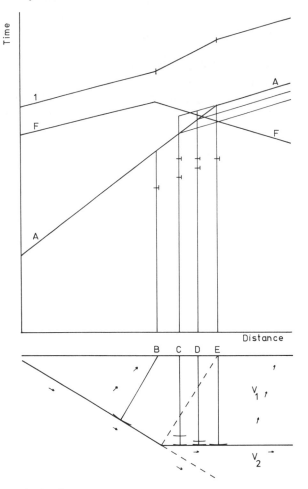

Fig. 4.27   Errors in depth.

Since this angle varies from $2\varphi_2$ at M to φ_2 at N, there is a continuous variation in the direction of the raypaths for the waves non-critically refracted into the V_1-layer. The diffraction zone at F constitutes an additional source of error. Improvements in the depth determination can be achieved if the different raypaths are considered and the longer travel paths from F and G are compensated by an appropriate reduction of the intercept times, a reduction based on the first estimate of the refractor configuration. The effect of the diffraction zone can be difficult to evaluate in practice. The errors in depth at points C and H are discussed below (see Figs 4.26 and 4.27). The depths obtained at D and E are slightly overestimated. At the wave front contacts on the ground, points B and K, the true depths are obtained.

Figures 4.26 and 4.27 depict the reverse and direct recording respectively for a partly sloping and partly horizontal structure. Points B and E refer to the wave front contacts on the ground surface. At these points the recorded arrival times yield the true depths of the refractor. Between B and E the depths are underestimated, as shown by the upper arcs below the depth determination points. The mean refractor velocity determination, curve 1, gives an indication of the positions of points B and E. The true intercept times in the critical region are obtained if, in the reverse and forward recording, the respective curve (F or A) is extended back to the impact points C and D and these constructed lines are used for the corrections. The upper short lines on the time axes through C and D designate the true intercept times. The depths obtained refer to different branches of the refractor depending on recording direction. In Fig. 4.26 the calculated depths have to be plotted perpendicularly to the sloping refractor section and in Fig. 4.27 vertically beneath the depth determination points. In fact, in the reverse recording the refractor configuration cannot be determined between point E and the point of dip change. If the dip change is combined with a low-velocity zone in the refractor, the error in depth increases.

4.1.4 The ABC method

This method belongs to the interpretation group where two emergence points on the refractor and a common point on the ground surface are considered. The time terms on which it is based were introduced by Edge and Laby in 1931. The interpretation technique was described by Heiland (1940) and Jakosky (1950) under the names 'ABC system' and 'method of differences' respectively. The method has since been further analysed and developed by Hagedoorn (1959), the plus part of the plus-minus method, and by Hawkins (1961), the reciprocal method. Since in the *Encyclopedic Dictionary of Exploration Geophysics* Sheriff (1973) has called the interpretation technique the 'ABC method', I prefer to use this name. Strictly, the ABC method is restricted by Sheriff to a use of the concept for weathering corrections but it is in agreement with the older denomination.

From the beginning the method was intended for depth determinations at the receiving stations but it can be applied to a number of other problems such as

1. Determination of total depths at geophone positions.
2. Determination of relative depths to be tied to depths obtained at impact points.
3. Correction of recorded arrival times and determination of intercept times at impact points.
4. Calculation of refractor velocities.
5. Supplying time terms for weathering corrections to be used in connection with reflection interpretations.

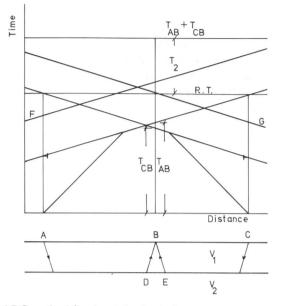

Fig. 4.28   The ABC method (horizontal refractor).

(a) Total depths

The method involves a summation of arrival times of critically refracted waves at a common surface point, B in Fig. 4.28, from two impact points A and C minus the reciprocal time (R.T.), i.e. the total time from impact point to impact point for waves propagated via the refractor. It is absolutely necessary that the times summed at B and the reciprocal time refer to the same refractor.

According to Fig. 4.28

$$T_{AB} + T_{CB} - T_{AC} = T_{DB} + T_{EB} - T_{DE}$$

$$= \frac{2h_1}{V_1 \cos i_{12}} - \frac{2h_1 \tan i_{12}}{V_2}$$

$$= \frac{2h_1}{V_1 \cos i_{12}} - \frac{2h_1 \sin^2 i_{12}}{V_1 \cos i_{12}}$$

$$= \frac{2h_1 \cos i_{12}}{V_1} \tag{4.32}$$

Thus

$$h_1 = \frac{(T_{AB} + T_{CB} - T_{AC})V_1}{2 \cos i_{12}} \tag{4.33}$$

or, in terms of velocities

$$h_1 = \frac{T_2 V_1 V_2}{2\sqrt{(V_2^2 - V_1^2)}}$$

(4.34)

The resulting time T_2 at B is the sum of the times associated with the upcoming rays. All factors outside the triangle DBE are eliminated. The time T_2 is composed of two identical delay times and since the same raypaths are involved, it is equivalent to the true intercept time, T_{2i}, for an impact point at B. The reversibility in intercept times obtained was previously mentioned in Section 4.1.1 (Fig. 4.1). Theoretically we ought to get the same depths and configuration of the refractor irrespective of whether we use the intercept times at impact points or at receiving stations. In practice, however, discrepancies are to be expected.

The identical intercept times – at the impact point or the geophone station – also imply that the depth determinations are affected by the same sources of error. Inherent in the ABC method are the assumptions that the refractor is plane between the emergence points D and E and that the refractor velocity is constant. Deviations from these assumptions cause errors. Note that the depth calculated at B depends on the local conditions at D and E.

Figure 4.29 refers to the case when the refractor is inclined. The derivation of the depth formula is restricted to the region DBG since terms outside the triangle are common to the travel times from the impact points to point B and in the reciprocal time but occur with different signs so that they cancel each other.

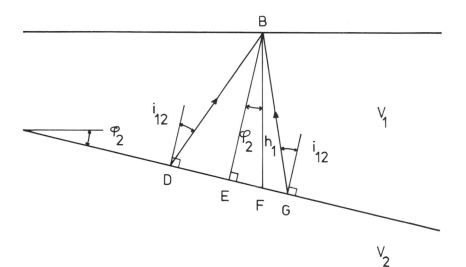

Fig. 4.29 The ABC method (dipping refractor).

Referring to Fig. 4.29

$$DB = GB = h_1 \cos\varphi_2 / \cos i_{12}$$

$$DE = EG = h_1 \cos\varphi_2 \tan i_{12}$$

and

$$h_1 = BF$$

The time T_2 at B is given by

$$T_2 = DB/V_1 - DE/V_2 + GB/V_1 - EG/V_2$$

in which there are two identical delay times. Thus

$$
\begin{aligned}
T_2 &= \frac{2h_1 \cos\varphi_2}{V_1 \cos i_{12}} - \frac{2h_1 \cos\varphi_2 \tan i_{12}}{V_2} \\[2mm]
&= \frac{2h_1 \cos\varphi_2}{V_1 \cos i_{12}} - \frac{2h_1 \cos\varphi_2 (1 - \cos^2 i_{12})}{V_1 \cos i_{12}} \\[2mm]
&= \frac{2h_1 \cos i_{12} \cos\varphi_2}{V_1}
\end{aligned}
\tag{4.35}
$$

which is the same as the intercept time obtained when the recorded arrival times are extrapolated to the energy source according to equation (4.5).

Solving for h_1, we obtain

$$h_1 = \frac{T_2 V_1}{2 \cos i_{12} \cos\varphi_2} \tag{4.36}$$

If the angle φ_2 is omitted in the calculations, the distance BE is obtained, perpendicular to the refractor. A tangential curve to the depth arcs at the receiving stations delineates the refractor surface.

The ABC method, summation of delay times, requires an overlap of recorded arrivals from the same layer. If in Fig. 4.28 the only available travel time curves were those from A and C, the method would be limited to the area around point B where there is an overlap of refractions from the second layer. However, the records from the offset points F and G, provided the arrival times are due to refraction from the V_2-layer, give a means to sum the delay times along the entire traverse in relation to the reciprocal time between A and C. In the figure, curves F and G have been displaced downwards and tied to the respective refractor velocity segments on the travel time curves from A and C, a procedure referred to as phantoming. A summation of the arrival times at every geophone station is thereby made possible. For the depth determinations, we must use the differences between the summed times and the reciprocal time for points A and C. Note that the construction with phantom arrivals is an application of the law of parallelism and, consequently, the intersections between the time axes through the impact points and the displaced curves yield the true intercepts at the impact points. These intercept times are equal to the time difference between $(T_{AB} + T_{CB})$ and the reciprocal time, R.T.

By summing the recorded arrival times along a profile, as indicated in Fig. 4.28, a so-called ABC-curve is obtained. This is the same as the plus curve in Hagedoorn's Plus-Minus method. If the ABC-curve forms a straight horizontal line (Fig. 4.28), the depth is constant provided there are no velocity changes in the overburden or in the refractor. Deviations from the horizontal indicate varying geological conditions along the profile in, for instance, depth, velocities and/or number of layers. This is illustrated in Fig. 4.30. The depth and the velocity V_1 in the overburden are assumed to be constant, while there is a velocity variation in the refractor, $V_{2'} < V_{2''} < V_2$.

The arrival times from the offset impact points A and D in Fig. 4.30 have been displaced as in Fig. 4.28 and tied to the travel time curves from B and C.

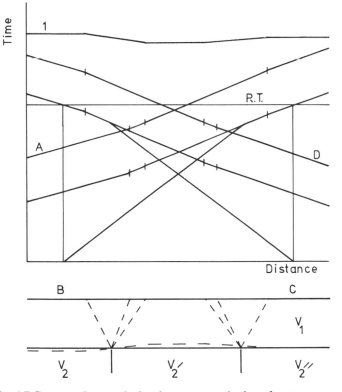

Fig. 4.30 ABC-curve when a velocity change occurs in the refractor.

There are no overlapping registrations of the refractor velocity from points B and C, a matter of no importance. The only demand is that the distance between B and C is sufficient to establish the reciprocal time, R.T. The missing arrival times are provided by the overlying curves A and D.

A summation of the recorded travel times does not result in this case in a horizontal, plane ABC-curve, curve 1 at the top of the time plot. The curve is partly horizontal, at the ends of the profile and in the centre of the $V_{2'}$ section. The transition above the velocity boundaries in the refractor coincides with the emergence at the ground surface of the wave front contacts and the outer limits of the diffraction zones. A simplified interpretation using an average refractor velocity of the area, assumed to be $V_{2''}$, leads to erroneous depths in the $\dot{V}_2$ and $V_{2'}$ parts of the profile. The refractor surface is then given by a dashed line. The errors in depth vary in this particular case between 5 and 7%. A computation using the velocities V_1 and V_2 results in an underestimation of the depth by about 10% in the $V_{2'}$ section. In the model shown the relative velocities V_1, $V_{2'}$, $V_{2''}$, and V_2 are 1, 1.75, 2.25, and 2.75 respectively.

It is advisable to be observant regarding depths calculated above low-velocity zones. They tend to be underestimated, a matter of importance in practice since such zones often constitute the most critical parts in engineering seismics. On the other hand, an interpretation based on a detailed refractor velocity analysis decreases the errors. In the example given in Fig. 4.30, the calculations have to be performed in such a way that the refractor velocity variation is considered, i.e. by considering the overburden velocity V_1 in relation to the appropriate refractor velocity. Some errors still remain in the transition sections above the velocity boundaries. These latter errors are less significant, however.

The field example of Fig. 4.31 was previously discussed in connection with the ABEM method, Fig. 4.20. The travel time curves from the end points, within the geophone spreads, and from the offset shots A and B are shown while the intermediate curves are omitted for the sake of clarity. The corrected intercept times at the shotpoints and corresponding depths (dashed lines) are, however, indicated. The summation of the recorded arrival times, curve 1, yields a mirror image of the bedrock surface since there is only one layer overlying the bedrock and the velocity in the saturated sand is very uniform, about 1700 m/s. The lower velocity in the bedrock (see Fig. 4.20) at the beginning of the profile slightly affects the ABC-curve (1).

Between 50 m and 75 m on the profile, curve 1 is rather horizontal, indicating small depth variations in this region. At the end of the profile, the curve shows that the depth increases. The dominant feature, however, is the steep slope of the curve in the beginning of the time plot. Depths calculated by the ABC method, according to the equation $h_1 = T_2 V_1/2\cos i_{12}$, are given by solid lines.

The intercept times at the shotpoints are almost of the same magnitude as the corresponding ones on the ABC-curve, which is to be expected since they

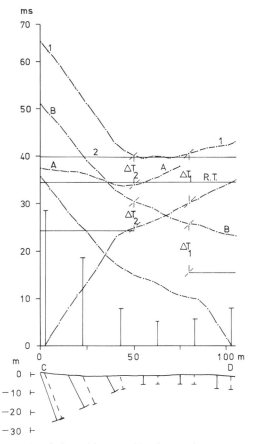

Fig. 4.31 Field example of Fig. 4.20 treated by the ABC method.

are based on the same concept. However, in practical refraction work greater discrepancies are to be expected in time reciprocity, in parallelism between velocity segments from the same layer, and between ABC-curves and intercept times at impact points due to, for instance, errors in time reading, in evaluation of impact instants or different raypaths.

In the construction of the ABC-curve in Fig. 4.31, the refractor velocity segments from C and D were completed by curves A and B, which were displaced and tied to curves C and D as demonstrated above. In order to avoid this tedious work, when computations are made graphically on a time–distance plot, a simpler and more rapid way is shown in the figure. Between 50 m and 80 m on the profile, the refractor curves from C and D were used for the construction of the ABC-curve. To complete the curve from 80 m to the end of the profile, the time difference ΔT_1 between curves C and curve 1 was plotted below curve B at the same station 80 m, and a horizontal reference line was drawn. Time differences between the reference line and curve B were then added to the arrival times on curve C at the corresponding geophone stations. A similar procedure was applied in the beginning of the profile. The time, ΔT_2, between curve 1 and curve A (at 50 m) was plotted below the corresponding arrival time on curve B and a horizontal reference line was drawn. The time differences between this line and curve B were then added to the travel times of curve A.

The application of the ABC concept to a three-layer case is presented in Fig. 4.32. It is assumed that the intermediate velocity V_2 is overlapping along the entire traverse. Curves B and F refer to offset impacts giving the velocity V_2. Curves A and G represent arrival times from the V_3-layer. In the usual manner the refractor velocity segments from points C and E have been completed by curves A and G. The upcoming rays at point H make the angles i_{23} and i_{13} with the interfaces, according to Section 3.2. Moreover, $\sin i_{23} = V_2/V_3$ (a critical angle) and $\sin i_{13} = V_1/V_3$. A summation of the V_3 travel times from C and E to H minus the reciprocal for C and E leaves a time value, T_3, represented by two delay times (identical) in the V_1- and V_2-layers respectively.

Therefore

$$T_3 = \frac{2h_2}{V_2 \cos i_{23}} - \frac{2h_2 \tan i_{23}}{V_3} + \frac{2h_1}{V_1 \cos i_{13}} - \frac{2h_1 \tan i_{13}}{V_2}$$

$$= \frac{2h_2}{V_2 \cos i_{23}} - \frac{2h_2 \sin^2 i_{23}}{V_2 \cos i_{23}} + \frac{2h_1}{V_1 \cos i_{13}} - \frac{2h_1 \sin^2 i_{13}}{V_1 \cos i_{13}}$$

$$= \frac{2h_2 \cos i_{23}}{V_2} + \frac{2h_1 \cos i_{13}}{V_1} \tag{4.37}$$

Solving for h_2, we obtain

$$h_2 = \frac{T_3 V_2}{2 \cos i_{23}} - \frac{V_2 h_1 \cos i_{13}}{V_1 \cos i_{23}} \tag{4.38}$$

which is the same as equation (3.15) since T_3 and T_{3i} are identical.

In the plot in Fig. 4.32, line 1 is the reciprocal time for the V_2-layer for waves from points C and D. The corresponding summation of delay times is given by line 2. The depth h_1 is calculated from the equation $h_1 = T_2 V_1/2 \cos i_{12}$. Lines 3 and 4 represent the reciprocal time for waves from C and E refracted via the V_3-layer and the summing of delay times respectively. As a comparison, the time difference between lines 2 and 1, the sum of delay times from the second layer, has been plotted from the reciprocal time (3), which gives the dashed line (5). The resulting time differences I_1 and I_2 are equal to I_1 and I_2 at the impact points.

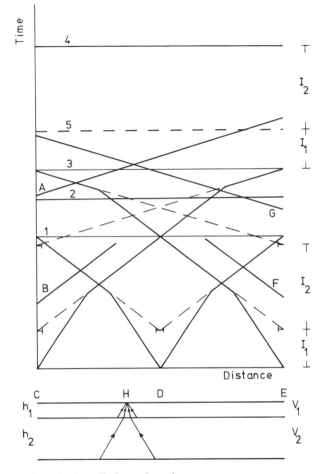

Fig. 4.32 ABC method applied to a three-layer case.

The general formulae for an arbitrary number of layers are as follows

$$h_{(n-1)} = \frac{T_n V_{(n-1)}}{2\cos i_{(n-1)n}} - \frac{V_{(n-1)}}{\cos i_{(n-1)n}} \sum_{v=1}^{v=n-2} \frac{h_v \cos i_{vn}}{V_v} \qquad (4.39)$$

or, expressed in velocities

$$h_{(n-1)} = \frac{T_n V_n V_{(n-1)}}{2\sqrt{(V_n^2 - V_{(n-1)}^2)}} - \frac{V_n V_{(n-1)}}{\sqrt{(V_n^2 - V_{(n-1)}^2)}} \sum_{v=1}^{v=n-2} h_v \sqrt{\left(\frac{1}{V_v^2} - \frac{1}{V_n^2}\right)} \qquad (4.40)$$

The general formulae for depth determination at the impact points (Section 3.3.1) are also valid when the delay times have been isolated at the receiving stations. The problem, however, is to get a complete coverage of the various velocities. Records of arrival times from the bottom refractor are almost

always obtainable, but the main difficulty lies in having continuous registrations of velocities from the intermediate layers without an excessive
recording with extremely small distances between the impact points. Records
containing second events would considerably improve the interpretation
situation and make it possible to have a very detailed picture of the interfaces
between the velocity layers.

Even if the overburden velocities at the geophone stations are not known in
detail, an ABC-curve can be used to delineate the bottom refractor. It goes
without saying that the curve must refer to the bottom refractor. Fictitious
velocity terms are computed yielding the total depth at the impact points.
These terms can then be used to calculate the depths at the geophones. A
multilayer case is then transformed into a simple two-layer case. If the
velocity terms at the impact points differ, a meaningful linear interpolation
can be made provided there is no sudden change in velocity.

(b) Relative depths

A simple and rapid way to detail a refractor surface is to make use of an
ABC-curve to calculate relative depths to be tied to those previously obtained
at the impact points. I refer to Fig. 4.31 for a demonstration of the interpretation technique. The shotpoint between 50 m and 75 m is selected as a base
for the depth determinations. At the intersection between a vertical axis
through the shotpoint and the ABC-curve, a horizontal reference line (2) is
drawn. Time differences between this line and the ABC-curve are used to
calculate the depth amounts to be added at the geophone stations – with
regard to their sign – to the base depth at the shotpoint. For the calculation of
the depth differences we can use either the ordinary equation $h_1 = T_2 V_1 / 2\cos i_{12}$ or a composite velocity term based on depth and time differences
between adjacent shotpoints. This term, or terms if the conditions vary along
the measuring line, can be regarded as a multiplication factor without a real
geological significance. In the depth calculation equation, $\Delta h_v = k \Delta T_v$, Δh_v is
the depth difference between the geophone station v and the selected shot-
point and ΔT_v is the corresponding time difference while k is the composite
velocity term.

Errors in depth are introduced if the velocity distribution in the refractor is
overlooked, particularly in the case of large velocity contrasts. Thus, in Fig.
4.30 the depression in the ABC-curve above the section with velocity $V_{2'}$ may
lead to erroneous depths. If the velocity variation is not considered, the
calculated depths will be overestimated or underestimated depending on the
selection of the base point for the relative depth determination.

It is vital when employing summation methods to establish the depths at the
receiving positions – total or relative – to have sufficient information about the
velocities in the layers overlying the refractor. A too-simplified field
measuring procedure with long distances between the impact points will, very
likely, lead to serious misinterpretations.

When determining relative depths it is not necessary to take into account the total summation time or the reciprocal time since only time variations of the ABC-curve are of interest. The summation of arrival times from the reverse refractor curves can be made from an arbitrarily chosen horizontal reference line. The ABC-curve is conveniently placed in an empty part of the time–distance graph.

Even if an ABC-curve is not used directly for calculations, it facilitates the analysis of the relationship between the travel time curves and the geology. Moreover, such a curve serves as an aid to check the depth calculations at the impact points and to estimate the refractor configuration between these points.

(c) ABC-curves used as correction means

The method can also be used to adjust irregular travel time curves in order to obtain true intercepts at the impact points.

For a two-layer case with parallel interfaces and constant velocities, the intercept time, which is the time at zero distance from the impact point, is

$$T_{2i} = \frac{2h_1 \cos i_{12}}{V_1} \tag{4.41}$$

The intercept time consists of two equal delay times. The depth h_1 is the only unknown quantity and the equation is solvable.

Assume now a two-layer case with constant velocities V_1 and V_2, but with varying depth of the refractor. The depths at the impact point A and at an arbitrarily chosen detector station B are h_{1A} and h_{1B}, $h_{1A} \neq h_{1B}$. The arrival time at B from the impact point A, is assumed to refer to the refractor velocity segment.

From the time–distance relation we obtain in this case the following apparent intercept time, when $x = 0$

$$T_{2iapp} = \frac{h_{1A} \cos i_{12}}{V_1} + \frac{h_{1B} \cos i_{12}}{V_1} \tag{4.42}$$

There are two unknowns, h_{1A} and h_{1B}, but only one equation and a unique solution of the depth at the impact point A (or at B) cannot be obtained. Because of the varying depth the intercept time is composed of two different delay times. The problem can be solved by making use of the summation times at A and B, obtained from an ABC-curve the arrival times of which originate from the refractor.

Half the summation time at A is

$$\frac{h_{1A} \cos i_{12}}{V_1}$$

and at B it is

$$\frac{h_{1B}\cos i_{12}}{V_1}$$

The time difference ΔT between A and B is

$$\frac{h_{1B}\cos i_{12}}{V_1} - \frac{h_{1A}\cos i_{12}}{V_1} \tag{4.43}$$

When the time difference ΔT in (4.43) is subtracted from (4.42), we get the expression

$$\frac{h_{1A}\cos i_{12}}{V_1} + \frac{h_{1B}\cos i_{12}}{V_1} - \left(\frac{h_{1B}\cos i_{12}}{V_1} - \frac{h_{1A}\cos i_{12}}{V_1}\right) = \frac{2h_{1A}\cos i_{12}}{V_1} \tag{4.44}$$

which is the true intercept at point A.

The correction technique is demonstrated graphically in Fig. 4.33. Points 1 and 2 represent the summation times from the reverse recording, i.e. the points lie on the ABC-curve for the refractor. The horizontal line drawn through point 1 gives the term $2\Delta T$, i.e. the difference between the intercept times at A and B. The travel time from A to B (3) is corrected by ΔT. In reality, several points on the refractor curve from A are corrected because the time readings may be scattered. Through the corrected time (4) a line with a slope equal to the reciprocal of the refractor velocity is drawn back to point A, giving the true intercept T_{2iA}.

In using the ABC method as a correction means, we are only interested in time differences between the ABC-curve and the horizontal reference line drawn through the intersection between the curve and the time axis at the particular impact point so that there is no need to take into account the total summation time or the reciprocal time. Since the method gives the double 'up-times', the time differences have to be divided by two to obtain the correction terms. This division by two can be carried out in constructing the ABC-curve itself.

In the case of an inclined refractor, the slope of the line through the corrected points is not $1/V_2$ but $\cos\varphi_2/V_2$, where the angle φ_2 is the slope of the refractor surface. This question will be discussed further in connection with the determination of refractor velocities in Section 4.2.1. However, the intercept time obtained is true, as it must be, since the correction line is drawn through the ABC-curve, vertically above the actual impact point. In Fig. 4.34, because of the sloping ground surface, the apparent velocities in the V_2- and V_3-layers are lower than the true velocities in the direct recording and higher in the reverse recording. The inclined ground surface also causes an apparent velocity in the top layer. As the real travel paths along the surface are longer than the corresponding horizontal distances, the recorded velocity V_1 in both directions is lower than the true velocity. Generally this source of error can be neglected but when the slope is appreciable, the recorded velocity has to be divided by the cosine of the slope angle.

The summation of the arrival times from the impact points A and G is represented by curve 1, which is the ABC-curve for the V_3-layer. Since there is coverage in both directions from the intermediate layer, V_2, it is feasible to construct an ABC-curve for this layer also. This is curve 2. In plotting the ABC-curves, the summation times have been divided by two. The recorded

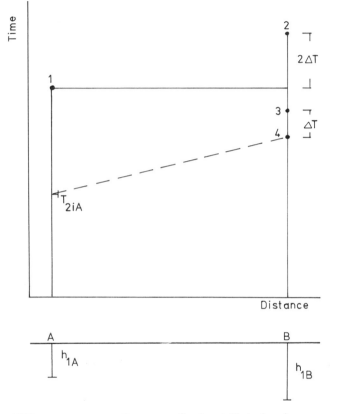

Fig. 4.33 ABC-curve as a correction means (horizontal interfaces).

arrival times from the impact points B, C and D have been corrected by means of the appropriate ABC-curve, curve 1 for the third layer and curve 2 for the second layer. The corrections are indicated by ΔT_1 and ΔT_2 on the time plot for the direct recording from point C.

If there is no complete overlapping of intermediate velocities, other solutions have to be resorted to. A lack of sufficient velocity coverage from the overburden is, in fact, the general case. An adjustment of apparent velocity segments with the aid of data from an underlying layer, using either

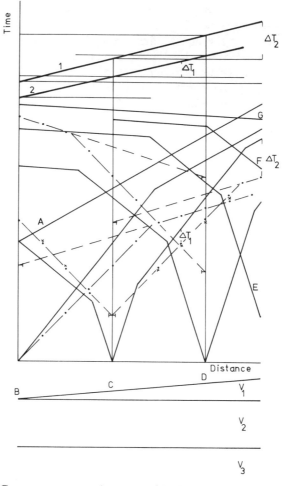

Fig. 4.34   ABC-curve as a correction means (sloping ground surface) (after Sjögren, 1980).

the ABC method or the ABEM method, is generally a less accurate solution, but it is sometimes the only possibility for making any sense of heavily distorted travel time curves. In Fig. 4.34 the adjusted arrival times (the V_2-layer) are shown by crosses when curve 1 is used as a means of correction. The corrected times, dots and crosses, differ slightly from each other. If the slopes of the velocity lines are based on the crosses, the computed velocities differ from the true velocities by less than 5% in this particular case. The residual error is, however, rather small compared with the deviation of 20–25% between the recorded and true velocities. The discrepancy between the corrected travel times, the dots and the crosses, depends on the non-parallelism between the ABC-curves 1 and 2. The curves diverge to the right

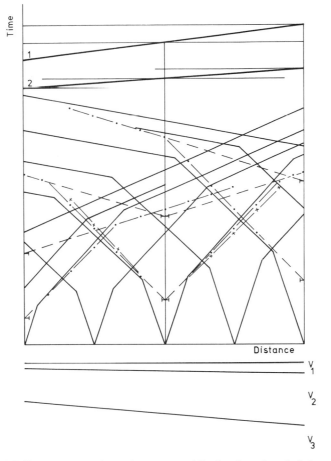

Fig. 4.35 ABC-curve as a correction means (dipping interfaces) (after Sjögren, 1980).

because of different raypaths through the upper medium. Assume that over a certain length Δx, the difference in depth is Δh_1. The corresponding time increments are then $2\Delta h_1 \cos i_{12}/V_1$ and $2\Delta h_1 \cos i_{13}/V_1$ on the ABC-curves 2 and 1 respectively. As $\cos i_{13}$ is greater than $\cos i_{12}$, the time increment ΔT_{13} in curve 1 is greater than the corresponding increment ΔT_{12} in 2. Therefore, the slope of curve 1 is steeper than that of curve 2.

Figure 4.35 illustrates a case where the ground surface is horizontal but the other interfaces are sloping to the right. The divergence between the ABC-curves 1 and 2 is more pronounced here than in the example of Fig. 4.34. Curve 2 for the V_2-layer is affected by the thickening of the V_1-layer, while in 1 there is a cumulative effect of the increase in thickness of the V_1-layer as well as of the V_2-layer. The differences in raypaths are less important in this case. The dots on the plot indicate the times when the apparent velocities have been corrected by the proper ABC-curve. The crosses refer to the case when the ABC-curve for the V_3-layer is used to adjust the recorded travel times of the V_2-layer. The intercept times obtained are true but there is a significant error in velocity. In such cases we have to evaluate the recorded and corrected velocities and their averages in both directions at the various impact points, and possibly velocity values from other seismic lines in the vicinity.

A combination of correction methods is given in the hypothetical example

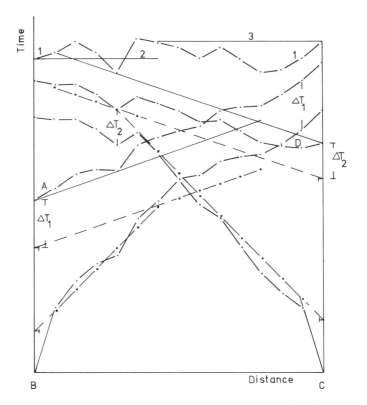

Fig. 4.36 Combination of correction methods (hypothetical data).

of Fig. 4.36. It is assumed that there is a coverage of arrival data from the second layer. Impact points between B and C, giving arrival times from the V_2-layer, are omitted for the sake of clarity. Velocity line segments from the bottom refractor (V_3) are corrected by either the law of parallelism, as shown by ΔT_1 and ΔT_2, or by the ABEM method, as shown by the dots. The second layer (V_2) is corrected by means of curve 1, the ABC-curve of the same layer. Note that curve 1 gives the double 'up-times' and the time differences between the curve and the reference lines 2 and 3 have to be divided by two before they are applied to the V_2 arrival times from B and C.

4.1.5 Hales' method

This interpretation method was introduced by F. W. Hales in 1958 for depth determinations. The method belongs to the group of methods where critically refracted rays diverging from a common point on the refractor are considered. The technique implies a summation of arrival times from a reverse recording at the emergence points on the ground surface for the rays from the common refractor point minus the reciprocal time between the impact points. Owing to the fact that the recorded travel times used for the interpretation refer to a common point on the refractor, the method is particularly suited for solving problems – depths and velocities – when the refractor surface is very irregular or at great depth. In such cases the interpretation techniques in which the arrival times for common surface points are used tend to give averaged results.

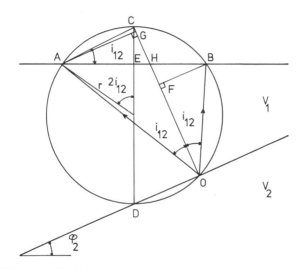

Fig. 4.37 Principle of Hales' method.

In Fig. 4.37 O is the emergence point on the refractor for the critical rays striking the ground surface at A and B. The problems to be solved are to fix the position of the points A and B associated with O and to determine the distance between the ground and the refractor at O.

For the explanation of the mathematical relations of the method, the triangle BOA is circumscribed by a circle. The line OC is the bisector of the angle AOB. The chords AC and CB are equal since their arcs subtend equal angles i_{12} on the circumference of the circle. A line from C through the centre of the circle bisects the distance AB. Since the angle COD is a right angle – OD being the tangent to the refractor at O – point D lies on the circle. The angle CAB is equal to i_{12}, because it is subtended by the same arc as the angle

COB. The lines AG and BF are drawn parallel to OD. Hence, they make an angle φ_2 with the horizontal and are normal to CO.

According to Fig. 4.37

$$AH + HB = AB$$

where

$$AH = AG/\cos\varphi_2$$

and

$$HB = BF/\cos\varphi_2$$

AG and BF can be expressed in terms of the raypaths AO and BO.

$$AG = AO\sin i_{12}$$

and

$$BF = BO\sin i_{12}$$

Thus

$$AB = \frac{AG + BF}{\cos\varphi_2}$$

$$= \frac{(AO + BO)\sin i_{12}}{\cos\varphi_2} \tag{4.45}$$

The sum of travel times from O to A and O to B is

$$T_2 = \frac{AO + BO}{V_1} \tag{4.46}$$

Replacing $(AO + BO)$ by $T_2 V_1$ in the expression for AB (4.45), we obtain

$$AB = \frac{T_2 V_1 \sin i_{12}}{\cos\varphi_2} \tag{4.47}$$

A circle with radius CO and centre at C will be tangent to the refractor at O. With the distance AB given by the equation above and the knowledge that C lies on the intersection between the perpendicular bisector of AB and a line from either A or B making the angle i_{12} with AB, the quantity to be determined in order to establish point O on the refractor is the distance CO. Now

$$\cos\varphi_2 = \frac{CO}{CD} = \frac{CO}{2r}$$

or

$$CO = 2r\cos\varphi_2 \tag{4.48}$$

where

$$r = \frac{AE}{\sin 2i_{12}}$$

and

$$2r = \frac{2AE}{\sin 2i_{12}} \qquad (4.49)$$

$$= \frac{AB}{\sin 2i_{12}}$$

Hence, using the expression $AB \cos\varphi_2 = T_2 V_1 \sin i_{12}$ according to (4.47), we obtain

$$CO = \frac{T_2 V_1}{2\cos i_{12}} \qquad (4.50)$$

The quantities needed for an interpretation using Hales' method are obtained by an ordinary reverse recording, such as that shown in Fig. 4.38 for a simple two-layer case with impact points at M and N. Curves 1 and 2 correspond to

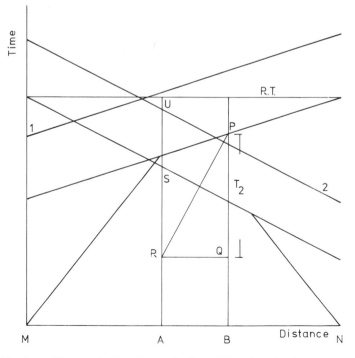

Fig. 4.38   Quantities needed for the application of Hales' method.

arrival times from the refractor recorded by offset impacts. The horizontal line R.T. is the reciprocal time for M and N.

According to the premises, the method is based on the summation of the recorded arrival times for the emergence points A and B for waves from the common refractor point minus the reciprocal time, i.e. $T_2 = T_{MB} + T_{NA} - T_{MN}$. If the travel time AS from N to A is plotted downwards from the reciprocal time R.T., i.e. AS = UR, and from R a line is drawn with a slope that is the inverse of $V_1 \sin i_{12}/\cos \varphi_2$ ($=AB/T_2$), it will intersect the refractor curve from M at P and the remaining time PQ is equal to T_2. Consequently, the distance RQ is equal to the distance between the points at which the waves emanating from the common refractor meet the ground surface. For the arbitrarily chosen point A, the associated point B has been established.

Hales demonstrated in his paper that the dip angle φ_2 can be disregarded in the interpretations, i.e. the assumption is made that the cosine of the angle is equal to unity. The effect of this simplification is that the point C in Fig. 4.37 moves towards the refractor by the same amount as the length of CO is reduced. The errors introduced cancel each other and only an insignificant source of error remains owing to the point O being moved along the refractor.

When only the curves from M and N are available, the depth calculation is limited to a central part of the traverse where the refractor velocity segments are overlapping. In order to complete the curves, the refractor velocity curves from the offset impact positions 1 and 2 are tied to the respective curves from M or N in the same manner as was previously mentioned in connection with the ABC method.

For manual interpretation work on a time–distance graph, Hales proposed to turn over one curve upside down placing the origin on the reciprocal time R.T. and plotting the arrival times downwards. In Fig. 4.39 the travel times of Fig. 4.38 are reproduced with the origin of curve N on R.T. Owing to the reciprocity, the time loop thus constructed has to close at M. The time AR in Fig. 4.39 is equal to US in Fig. 4.38, i.e. the remaining reciprocal time that has to be subtracted from the time BP on the forward refractor curve to fulfil the condition that $T_2 = T_{MB} + T_{NA} - T_{MN}$ (see Fig. 4.38). The distance between A and B, and therefore the position of P, are not the same in Figs 4.38 and 4.39, since for the slope of the line RP in Fig. 4.39 the simplified term $V_1 \sin i_{12}$ is used, i.e. the dip angle φ_2 is disregarded. A series of slope lines can be drawn between the two limbs of the time loop. For the sake of clarity, only three lines are shown in the figure. The time T_2 to be used in the depth formula (4.50) refers to the distance between Q and P. The centre points C, C', and C'' lie on the mid-points of respective AB distances along lines making the angle i_{12} with the horizontal. From the centre points short arcs of radii equal to the distances to the refractor are drawn. An envelope to these arcs maps the refractor shape. At both ends of the traverse there are parts of the approximate length $h_1 \tan i_{12}$ where the method cannot yield the depths of the refractor. If the continuous profiling procedure has been employed, the

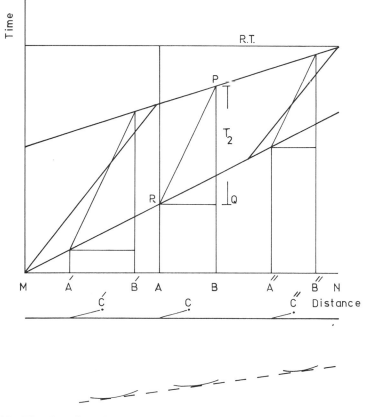

Fig. 4.39 Time loop in Hales' method.

various refractor curves from the geophone spreads can be connected to cover the entire profile length.

Figure 4.40 shows a part of a profile located in a water-covered area with water depths of 21–22 m. During the measurements the geophones and the shots were placed on the bottom. Geophone and shot distances are 5 and 25 m respectively. The travel time curves from the shotpoints are omitted and only the appropriate part of the time loop is given on the time plot. The bedrock velocity is about 5500 m/s, except for a velocity change between 180 and 200 m in the profile where the rock velocity is 3500 m/s. For the overburden a fictitious velocity term, 1850 m/s, is used. The travel time curves from the shotpoints indicated two velocities in the overburden, namely 1600 m/s and 2000 m/s. Before Hales' method can be applied to the problem a mean overburden velocity has to be computed, the aforementioned fictitious velocity term. The previously calculated depths and corresponding intercept times are used to obtain the mean velocity 1850 m/s. By this operation a three-layer case has been reduced to a two-layer case.

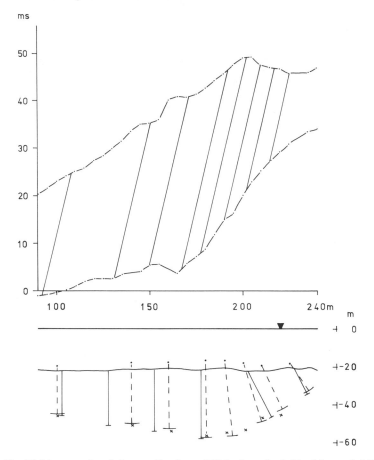

Fig. 4.40 Field example of the application of Hales' method (Archives of ABEM Company, Sweden).

Some of the slope lines are presented on the time–distance graph. The inclination, about 600 m/s, of these lines is obtained from the velocities 1850 and 5500 m/s applied to the term $V_1 \sin i_{12}$ (V_1^2/V_2). The radii of the arcs to be drawn from the centre points are calculated using the term $V_1 T_2/2\cos i_{12}$, i.e. $980 T_2$.

The calculated depths when using Hales' method are given by dashed lines. The solid lines refer to depths determined at the shotpoints. The crosses indicate depths obtained by drilling to bedrock or to other solid bottom material. The drilling was not extended into the rock. In the deepest part of the cross-section, at and immediately to the left of 200 m in the profile, the seismically determined depths are greater than those obtained by drilling. It is likely that the drilling has stopped on boulders without reaching the bedrock surface. Moreover, it may be assumed that the bedrock surface here is still deeper. In a depression in the bedrock, particularly in association with a low-velocity zone, seismics tends to underestimate the depth.

Since a lot of investigations are carried out in terrain with sloping ground surfaces it is of interest to extend Hales' method to include cases when the refractor as well as the ground are inclined. Near surface corrections may also result in non-horizontal reference lines. Moreover, when the method is used for an evaluation of refractor velocities, a neglect of dipping interfaces can cause errors.

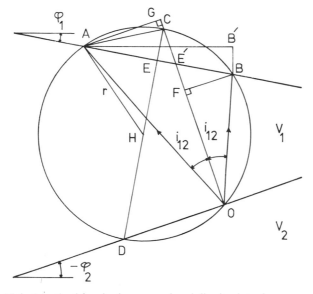

Fig. 4.41 Hales' method for sloping ground and dipping interface.

In Fig. 4.41 A and B are the points where the rays from the refractor point O strike the ground. CO is perpendicular to the refractor. Arc AC = arc CB (angles subtended are equal). Diameter CD bisects AB. COD is a right-angled triangle. $\angle$CAB = $\angle$COB = i_{12} (subtended by the same arc). AG and BF $\perp$ chord CO, therefore AG $\parallel$BF$\parallel$ refractor or to the tangent to the refractor at O. $\angle$GAB = $\angle$FBA = $\varphi_1 - \varphi_2$ = $\angle$DCO. Note that an angle in the first quadrant (φ_2 in this case) is given a minus sign and in the fourth quadrant a plus sign (φ_1).

Since the ground is sloping the distance AB' has to be determined. B' is the intersection between a horizontal line through A and a vertical line through B.

$AE' + E'B = AB$

$AG = AO \sin i_{12}$

$BF = OB \sin i_{12}$

$$AE' = AG/\cos(\varphi_1 - \varphi_2)$$

$$= AO\sin i_{12}/\cos(\varphi_1 - \varphi_2)$$

$$E'B = BF/\cos(\varphi_1 - \varphi_2)$$

$$= OB\sin i_{12}/\cos(\varphi_1 - \varphi_2)$$

Thus

$$AB = \frac{(AO + OB)\sin i_{12}}{\cos(\varphi_1 - \varphi_2)} \tag{4.51}$$

The horizontal distance AB' will then be

$$AB' = \frac{(AO + OB)\sin i_{12}\cos\varphi_1}{\cos(\varphi_1 - \varphi_2)} \tag{4.52}$$

Replacing $(AO + OB)$ by $T_2 V_1$, we get

$$AB' = \frac{T_2 V_1 \sin i_{12}\cos\varphi_1}{\cos(\varphi_1 - \varphi_2)} \tag{4.53}$$

The next step is to determine CO. The diameter of the circle = $2r$. According to the figure

$$\cos(\varphi_1 - \varphi_2) = CO/CD = CO/2r \tag{4.54}$$

and

$$2r = CO/\cos(\varphi_1 - \varphi_2)$$

From the triangle AEH

$$r = AE/\sin 2i_{12}$$

Since $2AE = AB$,

$$2r = AB/2\sin i_{12}\cos i_{12}$$

$$= CO/\cos(\varphi_1 - \varphi_2)$$

from which we obtain

$$CO = \frac{AB\cos(\varphi_1 - \varphi_2)}{2\sin i_{12}\cos i_{12}} \tag{4.55}$$

Replacing AB by $(AO + OB)\sin i_{12}/\cos(\varphi_1 - \varphi_2)$

$$CO = \frac{(AO + OB)}{2\cos i_{12}}$$

or

$$CO = \frac{T_2 V_1}{2\cos i_{12}} \tag{4.56}$$

which is the same expression as for a horizontal ground surface. When drawing the slope lines, the same simplification can be made as for the case with a horizontal ground surface, i.e. $AB'/T_2 = V_1 \sin i_{12}$. However, as mentioned above, the complete expression has to be used for the determination of the inclination of the slope lines when the method is used for refractor velocity analysis.

4.1.6 Different raypaths considered

As mentioned previously, interpretation methods can be divided into two groups depending on the raypath arrangement considered, namely times recorded at a common surface point or times referring to a common refractor point. The methods belonging to the former group yield an average picture of the subsurface conditions close to the impact or receiving point while the methods of the latter group can, under favourable conditions, make it possible to pin-point more accurately the structural details at relatively great depths. For smaller depths to the refractor or when the layer interfaces are fairly plane there is no appreciable difference between the methods. It may seem obvious that methods based on a definite refractor point are best suited to solve our problems but, unfortunately, varying conditions close to the ground surface may hamper or even make it impossible to apply them. The common near-surface corrections – recommended in the literature – are generally of little use in shallow refraction seismics since velocity and thickness changes in the often dry and loose top layers affect the recorded total times considerably, causing errors that are difficult to evaluate with the desired accuracy. This had already been pointed out by Domzalski in 1956.

The interpretation method to be employed depends on the type of project, the geological conditions and the measurement data available. Sometimes, in order to solve intricate problems, it is advisable to use a combination of various methods.

For an evaluation of the two raypath groups, a brief analysis of results obtained by the ABC method and Hales' method is given below. Some techniques – curve displacement and use of second events – to improve the interpretation results will also be discussed. The sources of error inherent in the various methods can be eliminated, or their influence on the results minimized, by technique modification in the field or in the interpretation work. Besides, even if we cannot reach a final, reliable solution in a particular case, knowledge of the errors to be expected is of value.

In the studies of sources of error intrinsic in the seismic method itself and in the various interpretation approaches, I have concentrated on problems connected with depressions in the refractor since they constitute the most critical parts of excavation works in rock. Misinterpretations may have serious consequences leading, for instance, to collapsing tunnels.

(a) Symmetrical depression – the ABC method

This case is shown in Fig. 4.42. The velocities V_1 and V_2 are assumed to be constant in the V-shaped depression. The travel time curves from A and E are given in the graph above the cross-section. Note that the curves represent apparent velocities from the down-dip and up-dip parts of the V_2-layer. The points B and D refer to the wave front contacts at the surface in the reverse and direct recording respectively. In a simplified model study these points are evident but generally they are not detectable in recorded, raw travel times owing to, for instance, time disturbances caused by near-surface irregularities. However, a refractor velocity analysis according to the 'mean-minus-T' method in Section 4.2.1 offers an aid to locate the emergence points of the wave front contacts. An interpretation according to this method, curve 1 in the figure, indicates a non-existing low-velocity zone in the central part of the depression. The limits of the false velocity zone coincide with points B and D. The distances BC and CD are equal to $h_c \tan i_{12}$, h_c being the vertical depth (CG) at C.

The ABC-curve, A′B′D′E′, gives the 'up-times' when the arrival times from A and E have been summed. In the regions AB and DE, the ABC method yields the true perpendicular distances to the refractor. Between B and D, however, the depth will be underestimated, as indicated by the upper arc beneath point C. Since the intercept times are the same in the region between B′ and D′, the calculated depth at C is equal to depths obtained at B and D. The error at C in this particular case – the model is based on $V_1 = 1500$ m/s, $V_2 = 5000$ m/s and $\varphi_2 = 33.7°$ – is about 34% of the total depth CG. The ordinary formula $h_1 = T_2 V_1 / 2 \cos i_{12}$ has been used for the depth determination.

The refractor section between I and K in Fig. 4.42 is unattainable by a conventional interpretation approach using the ABC method. The size of the shadow zone and consequently the error in depth depend on the magnitude of the dip angle φ_2 and the ratio between V_1 and V_2.

Since the calculated depth at C is the same as that at D, it can easily be derived from the figure. $\angle CGD = i_{12}$, $\angle GDP = \varphi_2$, $\angle PDK = i_{12}$, $h_c = CG$ and $GD = h_c / \cos i_{12}$.

From the triangle GDK we get the relation

$$DK = GD \cos (\varphi_2 + i_{12})$$

$$= \frac{h_c \cos (\varphi_2 + i_{12})}{\cos i_{12}} \tag{4.57}$$

which is also the depth that can be obtained at point C by the usual approach to the problem. The relation (4.57) is used below to evaluate the error in depth as a function of variations in dips and velocities.

In Fig. 4.43 the percentage errors in depth are plotted for varying dips and velocity ratios for two-layer cases. The numbers 0.1, 0.2, etc. represent the

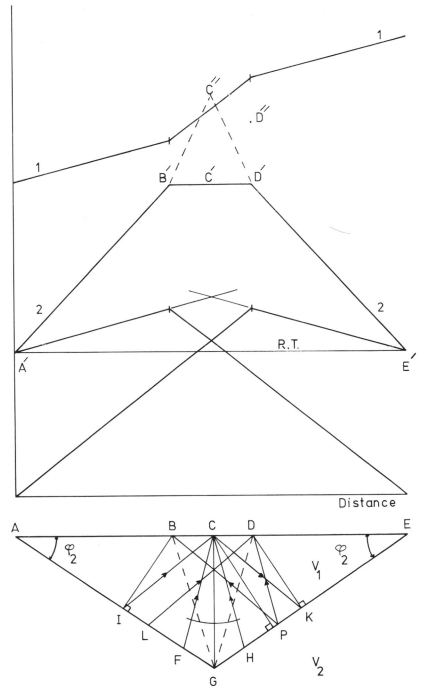

Fig. 4.42 ABC method applied to a symmetrical depression.

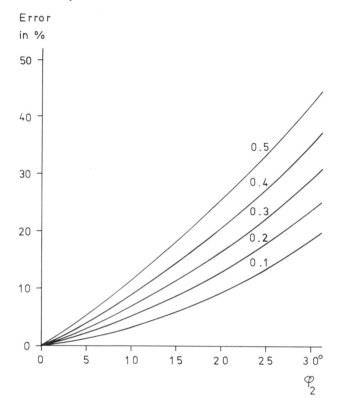

Fig. 4.43   Percentage errors due to dip using the ABC method.

ratio V_1/V_2. Note that the calculated errors are, in a sense, maximum errors. In practice, a depression is generally rounded rather than pointed as in Fig. 4.42 and the errors are smaller. However, we ought to reckon with a considerable error in the case of a pronounced depression in the refractor, especially when the velocity contrast is small.

There are, however, some simple means for improving the interpretation results in Fig. 4.42 and to evaluate the possible errors in the depth determinations. A cross-section based on the calculated depths at B and D yields an approximate value of the dip angle φ_2, the cosine of which can be used to estimate the vertical depths at B and D. The assumption involved is that the dips are constant in the depression. A practical way to get a better estimate of the depth is to place an extra profile in the vicinity of what is supposed to be the deepest part of the depression in the presumed direction of the depression. The depths obtained in a cross profile at point C in Fig. 4.42 will refer to distances perpendicular to the flanks of the depression, indicated by the line from C normal to GE. The shadow zone is reduced and consequently also the error in depth to be about 16% of the true depth CG.

(i) Modified ABC method

One assumption inherent in the ABC method and related methods is that the refractor is plane between the two points on the refractor from which the rays are critically refracted to the common point on the ground surface. Obviously, this is not the case in Fig. 4.42 for point C. There is, however, a simple way to evaluate the maximum depth in a depression. The derivation of the formula for a modified ABC method is as follows.

From Fig. 4.42

$$T_{AC} + T_{EC} - T_{AE} = T_{IC} + T_{KC} - T_{IG} - T_{GK} = T_2$$

Since $T_{IC} = T_{KC}$ and $T_{IG} = T_{GK}$, the equation becomes

$$T_2 = 2T_{IC} - 2T_{IG}$$

The law of sines applied to the triangle CIG, where $\angle CIG = 90° - i_{12}$, $\angle CGI = \varphi_2 + i_{12}$ and $\angle IGC = 90° - \varphi_2$, gives the following relations

$$IC = h_c \cos\varphi_2 / \cos i_{12}$$

and

$$IG = h_c \sin(\varphi_2 + i_{12}) / \cos i_{12}$$

Thus

$$T_2 = \frac{2h_c \cos\varphi_2}{V_1 \cos i_{12}} - \frac{2h_c \sin(\varphi_2 + i_{12})}{V_2 \cos i_{12}} \tag{4.58}$$

Replacing V_2 by $V_1 / \sin i_{12}$ and simplifying, the relation becomes

$$T_2 = \frac{2h_c \cos(\varphi_2 + i_{12})}{V_1} \tag{4.59}$$

from which we obtain

$$h_c = \frac{T_2 V_1}{2\cos(\varphi_2 + i_{12})} \tag{4.60}$$

The equation contains two unknowns, h_c and the dip angle φ_2. The interpretation has to be carried out in stages. An approximate value of φ_2 can be obtained from the first depth calculation using the ordinary procedure. The estimate of the angle φ_2 can then be used in the modified Equation (4.60). The calculation is repeated until the depth obtained remains stable. The interpretation can also commence with a dip angle based on an assumed depth. A calculation based on an overestimate of the depth leads rapidly to the true depth.

The source of error that causes the depth to be underestimated when the ordinary interpretation technique is employed can be identified from Equation (4.58).

$$T_2 = \frac{2h_c\cos\varphi_2}{V_1\cos i_{12}} - \frac{2h_c\sin(\varphi_2+i_{12})}{V_2\cos i_{12}}$$

$$= \frac{2h_c\cos\varphi_2}{V_1\cos i_{12}} - \frac{2h_c\sin\varphi_2\cos i_{12}}{V_2\cos i_{12}} - \frac{2h_c\cos\varphi_2\sin^2 i_{12}}{V_1\cos i_{12}}$$

$$= \frac{2h_c\cos i_{12}\cos\varphi_2}{V_1} - \frac{2h_c\sin\varphi_2}{V_2} \qquad (4.61)$$

Since the first term on the right-hand side of the equation yields the true intercept time for a dipping refractor, the second term represents the error. Note that the same source of error was encountered when depth determination at impact points was analysed in Section 4.1.3, Equation (4.31), for a similar structure.

(ii) Curve displacement
One way to improve the depth analysis for a depression is to displace the curves in relation to each other, for instance the curve E in Fig. 4.42 to the right and to carry out a renewed summation of the arrival times. A displacement of the curve by $2h_c\tan i_{12}$ is equal to the distance between B and D, where the maximum intercept time is to be obtained. An indication of the distance BD may be had from the form of the recorded curves and from the determination of the refractor velocity as the distance between the points at which the velocity changes. If the maximum displacement can be established from such indications, the corresponding intercept time can be used in the ordinary depth formula $h_1 = T_2V_1/2\cos i_{12}$. For the example in Fig. 4.42 the remaining error is about 9% when this interpretation technique is applied to the problem. The interpretation can also be carried out as a successive curve displacement. If the depth at C when undisplaced curves have been used is designated h_{c1}, the displacement is $2h_{c1}\tan i_{12}$ and the new depth is calculated according to the equation above. This procedure can be repeated until a maximum depth is obtained. A displacement of curve E so that the arrival time at B happens to lie to the right of D on the curve from A leads to a decrease in the intercept time because of the negative slope of the up-dip refractor velocity segments.

The displacement technique described above improves the results of the depth determinations, but there is still a remaining error when the ABC method is used in the usual manner, because the appropriate travel ways are not considered. This question will be treated below.

If in Fig. 4.42 the curve from E is displaced $2h_c\tan i_{12}$ to the right, the travel time at B will coincide with that at D in curve A. The sum of the arrival times is denoted by point D''. Mathematically this can be expressed in the following way

$$T_{AD} + T_{EB} - T_{AE} = T_2 = T_{LD} + T_{PB} - T_{LG} - T_{GP}$$

where

$$T_{LD} = T_{PB}$$

and

$$T_{LG} = T_{GP}$$

When the law of sines is applied to the triangle LDG, where $\angle GDL = \varphi_2$, $\angle DLG = 90° - i_{12}$, $\angle LGD = 90° + (i_{12} - \varphi_2)$ and $GD = h_c/\cos i_{12}$, the following relations are obtained

$$LD = h_c \cos (i_{12} - \varphi_2)/\cos^2 i_{12}$$

and

$$LG = h_c \sin \varphi_2/\cos^2 i_{12}$$

Then the intercept time T_2 can be expressed in the form

$$T_2 = \frac{2h_c \cos (i_{12} - \varphi_2)}{V_1 \cos^2 i_{12}} - \frac{2h_c \sin \varphi_2}{V_2 \cos^2 i_{12}} \qquad (4.62)$$

Replacing V_2 by $V_1/\sin i_{12}$, the equation becomes

$$T_2 = \frac{2h_c \cos \varphi_2}{V_1 \cos i_{12}} \qquad (4.63)$$

or

$$h_c = \frac{T_2 V_1 \cos i_{12}}{2 \cos \varphi_2} \qquad (4.64)$$

When using this equation, a first estimate of the depth or an assumed depth can form the basis for an evaluation of the dip angle and for the magnitude of the displacement. The procedure can be repeated until the difference between assumed and calculated depths is negligible. If the technique with an assumed depth is used, it should be noted that a deliberate overestimate of the depth solves the problem most rapidly.

The general depth determination formula for displaced curves is

$$h_1 = \frac{T_2 V_1 \cos i_{12}}{2} \qquad (4.65)$$

when the dip angle $\varphi_2 = 0$.

Referring to Fig. 4.44

$$T_{AD} + T_{BC} - T_{AB} = T_{OD} + T_{OC} = T_2 = \frac{2h_1}{V_1 \cos i_{12}}$$

which gives the equation above.

For depth calculations, one curve is displaced $2h_1 \tan i_{12}$ and the arrival times are added. This procedure is indicated in Fig. 4.44. The arrival time C′ is moved to the right to have the same horizontal position as D′. The sum of the arrival times minus the reciprocal time R.T. gives the time T_2. The depth obtained has then to be plotted a distance $h_1 \tan i_{12}$ to the left of point D.

The interpretation technique can be used to study the depths of local features in the refractor configuration, for instance buried channels and ridges. The magnitude of the displacement can be based on the preliminary depth calculations at the impact points or on a conventional use of the ABC method. A successive displacement technique can also be used to study the depths obtained with an increasing magnitude of displacement. Relative depths, to be tied to depths obtained at the impact points, can be calculated using the formula above.

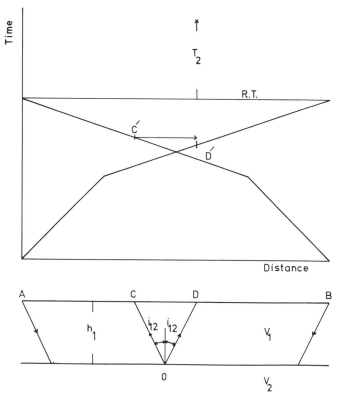

Fig. 4.44 Displacement procedure.

Depth determinations according to displacement procedures ought to be checked by dips obtained from the recorded refractor curves and possibly by features of significance in the curves and the refractor velocity determinations to find out if there are any disagreements.

For a further study of displacement techniques the reader is referred to L. C. Pakiser and R. A. Black (1957) and D. Palmer (1980).

(iii) Construction lines
The modified and displacement techniques described above are based on the recorded first arrival times. In Fig. 4.42 the up-dip velocity segments from A and E have been extended to the centre of the depression. These construction lines correspond to second arrivals. A summation of these real or imaginary times gives a new intercept time, C'' on the plot. When using this intercept for calculations, the ABC concept has to be modified since other raypaths are involved. In the direct recording the waves have to travel via point H to reach the ground at point C and in the reverse recording via F to C.

Since the triangles CGF and CHG are equal, it is sufficient to study the conditions in one wave-propagation direction. In the triangle CHG, $\angle HGC = 90° - \varphi_2$, $\angle CHG = 90° + i_{12}$, $\angle GCH = \varphi_2 - i_{12}$, and $h_c = CG$.

The law of sines applied to the triangle CHG gives the relations

$$GH = h_c \sin(\varphi_2 - i_{12})/\cos i_{12} = FG$$

and

$$HC = h_c \cos \varphi_2 / \cos i_{12} = FC$$

The travel times from A to C and from E to C are added and then diminished by the time for the travel path AGE, which is the reciprocal time.

$$T_{AC} + T_{EC} - T_{AE} = T_2 = T_{AF} + T_{FG} + T_{GH} + T_{HC} + T_{EH} + T_{GH} + T_{FG} + T_{FC}$$

$$- T_{AF} - T_{FG} - T_{GH} - T_{EH}$$

$$= T_{HC} + T_{FC} + T_{GH} + T_{FG}$$

Because of the symmetry

$$T_2 = 2T_{HC} + 2T_{GH} = \frac{2h_c \cos \varphi_2}{V_1 \cos i_{12}} + \frac{2h_c \sin(\varphi_2 - i_{12})}{V_2 \cos i_{12}}$$

When V_2 is replaced by $V_1/\sin i_{12}$

$$T_2 = \frac{2h_c}{V_1}(\sin \varphi_2 \sin i_{12} + \cos \varphi_2 \cos i_{12}) = \frac{2h_c}{V_1} \cos(\varphi_2 - i_{12}) \qquad (4.66)$$

or

$$h_c = \frac{T_2 V_1}{2\cos(\varphi_2 - i_{12})} \qquad (4.67)$$

The ordinary expression $T_2 V_1/2\cos i_{12}$ can be used in combination with the intercept time C'' to get an approximate value for the depth at point C. The dip angle then obtained can be used in the Equation (4.67) for a more accurate estimate of the depth h_c and if desired the calculations can be repeated. Note that when φ_2 is less than i_{12}, no second events from the up-dip velocity segments can reach point C. The turning-point is when φ_2 is equal to i_{12}.

(b) Symmetrical depression – Hales' method

Hales' method applied to the structure of Fig. 4.42 also results in an under-estimate of the depths in the central part of the structure. Because of the different travel paths considered, it is, however, possible to penetrate deeper into the depression by Hales' method than by the ABC method.

The structure and time–distance graph shown in Fig. 4.45 correspond to those of Fig. 4.42. The time loop is closed between A and E'. The slope line 1 is based on the simplified term $V_1 \sin i_{12}$. Owing to the neglect of the dip angle φ_2, the centre point lies to the right of C. The depth is plotted perpendicularly to the refractor (GE). When the exact expression $V_1 \sin i_{12}/\cos \varphi_2$ is used to

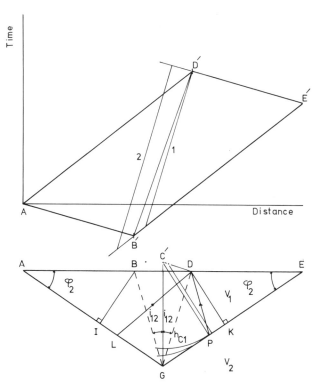

Fig. 4.45   Hales' method applied to a symmetrical depression.

calculate the inclination of the slope lines, the line from B′ intersects the direct curve at D′. The centre point of this line has the same horizontal position as C. The depth refers to the points L and P on the refractor. The error in depth is now about 19%, compared with 34% when the ABC method is applied to the problem. If the effect of the dip is not considered, the error is about 22% of the true depth CG. If second arrivals have been observed, they can be used to delineate the refractor configuration more accurately. Slope line 2 is drawn between 'outer' parts of the time loop and the depth obtained refers to the right-hand flank of the depression, shown by the lower arc.

The difference in depth obtained depends on the size of the shadow zone in the refractor. When the ABC method is used the maximum depths – based on the first arrivals – obtainable between B and D are the identical distances BI and DK, which means that the shadow zone lies between I and K. Hales' method shrinks the zone so that it lies between L and P or in the vicinity of these points if the influence of the dip is neglected. L and P are the points where the waves leave the refractor to strike at B and D. This condition enables us to calculate the errors to be expected in a V-shaped depression with varying velocities and dips.

Referring to Fig. 4.45

$$DP = h_c \cos(\varphi_2 + i_{12})/\cos^2 i_{12}$$

The law of sines applied to the triangle C′DP, where $\angle DC'P = 90° - \varphi_2 - i_{12}$ and $\angle PDC' = 90° + \varphi_2$, gives

$$C'P = \frac{DP \cos\varphi_2}{\cos(\varphi_2 + i_{12})}$$

Thus

$$C'P = \frac{h_c \cos\varphi_2}{\cos^2 i_{12}}$$

The apparent depth h_{c1} below C will be obtained by subtracting $h_c \tan^2 i_{12}$ from C′P so that

$$h_{c1} = \frac{h_c(\cos\varphi_2 - \sin^2 i_{12})}{\cos^2 i_{12}} \tag{4.68}$$

This relation has been used to compute errors in depth (Fig. 4.46) as a function of the dip angle and the velocity ratio V_1/V_2. The numbers 0.1, 0.3, and 0.5 in the figure designate the velocity ratios. An interpretation made by disregarding the dip angle increases the errors in the diagram by a few per cent. In any case, the errors are considerably less than those shown in Fig. 4.43 for the ABC method.

(c) U-shaped depression

The recorded first arrival times in Fig. 4.42 do not give any indication whether the depression in the V_2-layer is V-shaped or less pointed. The latter case will

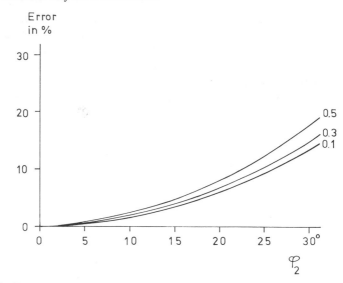

Fig. 4.46   Percentage errors due to dip using Hales' method.

be discussed by reference to Fig. 4.47 where a part of the refractor is horizontal in the centre of the cross-section. Other pertinent data are the same as for Fig. 4.42.

Since the travel paths in the refractor are shorter here than in Fig. 4.42, the waves return earlier to the ground and consequently the distance between the emergence points B and D is shorter. The shadow zone is smaller than in Fig. 4.42.

When the ABC method is used for the depth calculations, the error is about 17% as seen from the upper arc below point C. In fact, the depth obtained is greater than in Fig. 4.42 when the calculations are based on the first arrivals in a conventional manner. This is a paradoxical situation in that an actual increase in depth below C results in a decrease in calculated depth. An application of Hales' method gives a difference of some few per cent between real and calculated depths, indicated by the lower arc. Interpretation techniques with curve displacement, constructions, etc. can result in an overestimate of the depth. On the other hand, the interpretation situation would be considerably improved if second events were available for an analysis of the refractor configuration.

Also in the case of a more gradual change in the shape of the refractor, the calculated depths tend to be underestimated. The depth (upper arc) at point C in Fig. 4.48 is obtained by the ABC method. This depth is equal to the perpendicular distances to the refractor at B and D since the intercept time is constant between these points in the ABC-curve (1). However, Hales' method yields an acceptable picture of the refractor relief as shown by the arc close to the refractor. The modified techniques proposed can improve the

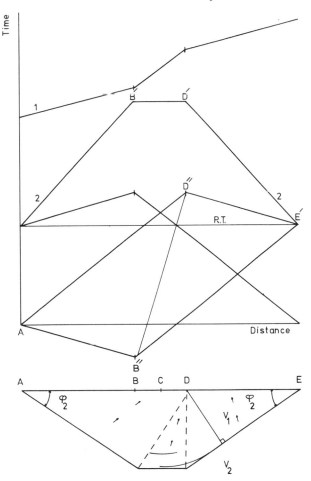

Fig. 4.47 Depth errors in a U-shaped depression.

interpretation results when the ABC method is used. Of interest is the fact that even if travel time curves are available from all parts of a depression, there is still a possibility that we underestimate the depths. In the case of even a slightly undulating refractor surface, a seismic survey tends to underestimate the depths in the troughs and overestimate the depths above the culminations.

(d) Low-velocity zone in a depression

Such a case is illustrated in Fig. 4.49. Except for the velocity change in the refractor the structures of Figs 4.49 and 4.42 are the same.

Because of the lower velocity in the refractor, the waves are delayed and return to the ground surface later than in the example given in Fig. 4.42,

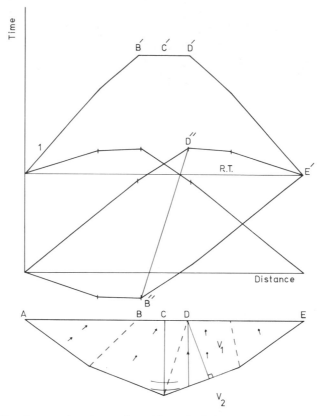

Fig. 4.48   Depth errors in a depression with gradual change.

causing a widening of the distance between the wave front contacts B and D. The shadow zone is therefore greater than in Fig. 4.42. The location of points B and D can be found from the mean refractor velocity curve (1). For an explanation see Section 4.2.1. The intercept times B' and D' on the ABC-curve (2) yield the true perpendicular distances to the refractor. This is also the depth obtained at point C, shown by the upper arc. The interpretation underestimates the depth by about 40%, compared with 34% for the structure in Fig. 4.42.

A construction of the up-dip velocity segments to point C, as proposed for the structure in Fig. 4.42, leads to an overestimation of the depth by about 6%. The construction assumes a constant refractor velocity along the entire traverse and the time delay in the lower velocity layer is not taken into account. The delay is $(FG + GH)/V_{2'} - (FG + GH)/V_2$. The other interpretation techniques proposed for the case of Fig. 4.42 give varying results. The equation $h_c = T_2 V_1 \cos i_{12}/2\cos \varphi_2$, (4.64), tends to slightly overestimate the depth, while a calculation based on the equation for the modified method

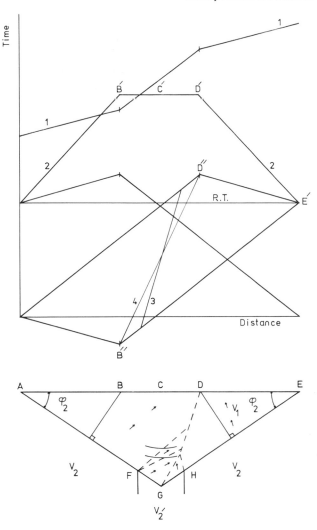

Fig. 4.49 Depth errors in a depression with low-velocity zone.

$h_c = T_2 V_1/2\cos(\varphi_2 + i_{12})$, (4.60), underestimates the depth. The reason is that the critical angle i_{12} refers to the velocities V_1 and V_2 and not to V_1 and $V_{2'}$.

Line 3 in the time loop shows the case when the inclination of the slope lines is based on the term $V_1 \sin i_{12}$. The corresponding depth obtained is shown by the lower arc beneath point C. The error in depth is about 30%. The error is less than with the ABC method but is considerably greater than the error in Fig. 4.45. In line 4 the lower velocity in the refractor is considered when evaluating the inclination of the slope lines. It is then possible to reduce the error to about 18% in the model study.

(e) Asymmetric depression

Figure 4.50 shows an asymmetric depression in the refractor so that $\varphi_2' > \varphi_2$. A considerable part of the refractor relief is unattainable in this case when using conventional interpretation methods. The shadow zone (the ABC method) is to be found between the feet of the perpendiculars from B and D to the respective flanks of the refractor. The upper arcs on the cross-section refer to calculations according to the ABC method. A significant feature is that the depth interpretation yields a distorted picture of the real refractor configuration. The calculated depths increase towards point B, while the actual maximum depth lies closer to point D. Note that in the ABC-curve the intercept times increase from D' to B'. The reason for the depth anomalies is to be found in the fact that the angle φ_2' is larger than φ_2. The error in depth at point C is about 35% when the ABC method is used.

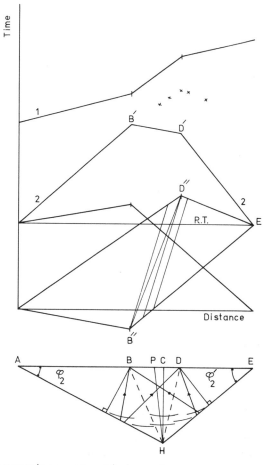

Fig. 4.50   Depth errors in an asymmetric depression.

Using Hales' method reduces the error to about 25% at point C. The lower arcs refer to Hales' method. We observe here the same tendency, namely that the calculated depths increase towards B. When the dips are considered in the calculation, as in drawing the slope line between B″ and D″, the error is about 20% at point C.

An estimate of a possible maximum depth can be made with the aid of the equation $h_c = T_2 V_1 \cos i_{12}/2\cos \varphi_2$ (the ABC method). The equation implies a symmetrical depression but nevertheless it can be used for a non-symmetrical shape of the refractor, for which a mean of the dip angles is sufficient. The interpretation can be made by either a successive displacement, the crosses on the graph, or by assuming an overestimated depth. For the displacement, the term $2h_{cv} \tan i_{12}$ can be used, despite the fact that the angle DHC and BHC are not equal to i_{12}, as for a symmetrical depression, but to $i_{12} - (\varphi_2' - \varphi_2)/2$ and $i_{12} + (\varphi_2' - \varphi_2)/2$ respectively. When the depth obtained is moved back half the displacement distance, the depth is plotted, not below C, but below P, the midpoint of BD. The point H for the maximum depth is displaced by $h_c \tan (\varphi_2' - \varphi_2)/2$ in relation to P.

If second events are detectable in the records, they and the ordinary formula $h_c = T_2 V_1/2\cos i_{12}$ may be used to improve the accuracy of the depth determinations. However, the correct equation for second events in the case of a non-symmetrical depression is $h_c = T_2 V_1/[\cos (\varphi_2 - i_{12}) + \cos (\varphi_2' - i_{12})]$.

(f) Faults

Figures 4.51–4.54 illustrate the typical travel time curves in the case of large scale faulting. The refractor velocity is assumed to be constant in Figs 4.51 and 4.52 while in Figs 4.53 and 4.54 there is a velocity change in the downthrown part of the fault, $V_{2'} < V_2$. The wave front pattern is rather complex, particularly for the structure with a velocity change in the refractor. Dominant wave front features are:

(a) The diffraction zone emanating from the edge of the fault in the direct recording.
(b) The vertical or almost vertical raypaths towards the upthrown section of the fault in the reverse recording.
(c) The retrograde direction of the waves from the edge in the reverse recording.

The low-velocity zone in the refractor causes a widening of the diffraction zone in the forward recording in Fig. 4.53 compared with the case in Fig. 4.51 where the waves from the downthrown refractor section take over more rapidly. In the reverse recordings the refractor velocity segments are horizontal above the edge on the upthrown side of the fault. With increasing distance from the edge to the left, the slope of the reverse curves asymptotically approaches a slope equal to the inverse of the true velocity V_2.

The mean refractor velocity determinations according to Section 4.2.1 are

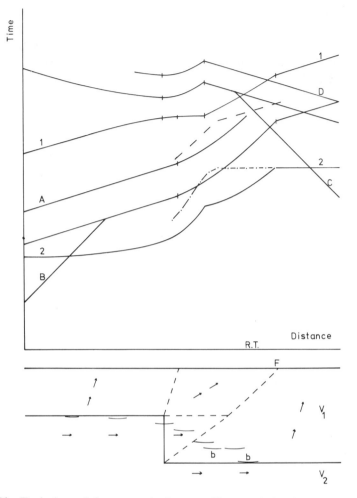

Fig. 4.51 Typical travel time curves in the case of large scale faulting.

given in curves 1 in Figs 4.51 and 4.53. The ABC-curves (2) are moved into the time–distance graph in order to keep the size of the figures small and the reciprocal times R.T. are placed on the zero time line.

The depth analysis (the ABC method) in Fig. 4.51 partly smooths out the prominent step in the refractor surface. On the upthrown part of the refractor the depths are overestimated because of the direction of the raypaths in the reverse recording. The error decreases to the left of the edge as the angle that the rays make with the refractor surface decreases. On the downthrown section, the true depth is obtained at point F which is the wave front contact on the ground. In Fig. 4.52 (Hales' method) the refractor relief can be traced

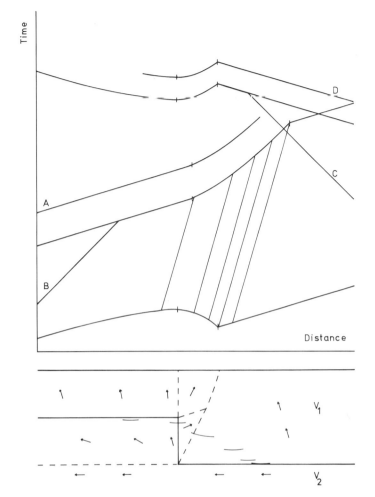

Fig. 4.52   Typical travel time curves in the case of large scale faulting.

closer to the fault. In general, the errors are less when Hales' method is
utilized.

Improvements in the depth calculations can be made by displacing the
curves or by means of second events, if available. The depths marked (b) in
Fig. 4.51 refer to a displacement of the curve from B towards the left as
indicated by the dashed line on the time plot. The new ABC-curve in this
section is given by the dash-dotted line tied to the original ABC-curve (2).
The displacement is based on the depth calculated at F and to the right of F. A
displacement of the curve from C above the upthrown part of the fault
decreases the error (the ABC method). If the project in question requires a

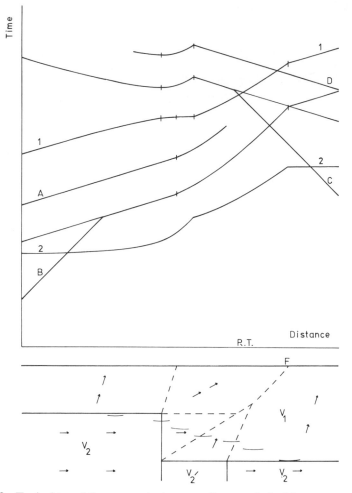

Fig. 4.53 Typical travel time curves in the case of large scale faulting.

more accurate estimate of the location of the fault, the broadside profiling technique may be employed, i.e. a series of impacts are made in a distant line parallel to the geophone line. Such measurements should be carried out on both sides of the geophone line in order to minimize the influence of varying refractor velocities. A low-velocity zone crossing the geophone line may otherwise heavily distort the registrations.

The presence of the low-velocity zone in Figs 4.53 and 4.54 increases the error in depth in the downthrown part of the fault. The distance from the edge of the fault to point F, where the true depth is obtainable, is also increased. Improvements in the depth calculations can be achieved by displacement technique or possibly by using second events.

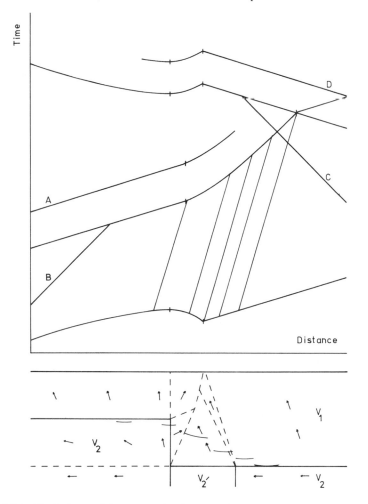

Fig. 4.54 Typical travel time curves in the case of large scale faulting.

(g) Ridges

The errors to be expected in the case of a ridge occurring in the refractor have been discussed in connection with Fig. 4.25. The ABC method gives the same results and therefore also the same errors. On the top of a ridge, the calculated depths tend to be greater than the true depths. However, Hales' method yields a more accurate picture of the refractor configuration. The same result can be obtained by using the displacement technique for the ABC method. A further improvement of the determination of the refractor shape can be achieved if the influence of the raypaths along the flanks of the ridge is considered.

4.1.7 Other methods

The interpretation methods treated in the previous sections of this chapter are
– in my opinion – particularly suited to solving shallow refraction problems
connected with greatly varying geological conditions. In other methods such
as, for instance, delay time methods and wave front methods, more assumptions are involved, decreasing the reliability of the results or making them
more laborious. Moreover, as these methods are thoroughly described in the
literature I shall restrict myself to referring only to some authors. For interpretation techniques using the delay time concept mention can be made of
Gardner (1939, 1967), Barthelmes (1946), Wyrobek (1956), Tarrant (1956),
Layat (1967) and Barry (1967). A comprehensive description of wave front
methods was given by Rockwell in 1967. The classic paper on this subject was
written by Thornburgh in 1930. Other papers of interest are those by Gardner
(1949), Baumgarte (1955), Meissner (1965) and Schenck (1967). The various
interpretation methods are also summarized by Dobrin (1976) and by
Telford, Geldart, Sheriff and Keys (1976).

4.2 REFRACTOR VELOCITY DETERMINATIONS

Two different methods for computing refractor velocities will be discussed
below. Both methods require direct and reverse recordings. The difference
between the techniques lies in the raypath arrangement considered. The
'mean-minus-T' method belongs to the group with a common surface point to
which the waves involved are critically refracted from two points on the
refractor. The other is Hales' method which is based, as mentioned earlier, on
the recording of waves critically refracted from a common refractor point.
The mean-minus-T method is useful for relatively shallow depths, particularly if there are large variations in velocities and thicknesses in the
uppermost layers, while Hales' method yields more reliable results in the case
of high-relief structures or for great depths to the refracting horizons. The
mean-minus-T method tends in such cases to give a general picture of the
velocity distribution or sometimes even erroneous results. On the other hand,
very irregular top-layer conditions for which corrections cannot be made with
the needed accuracy may make an application of Hales' method impossible.

A convenient interpretation procedure is to start with the mean-minus-T
method to get the general velocity picture. Later a more detailed velocity
analysis may be made, if necessary, by Hales' method. Moreover, the general
velocities are needed for the preliminary depth determinations. Depth and
refractor velocity determinations are, strictly speaking, inseparable. In order
to obtain accurate depths we need an accurate refractor velocity analysis and,
conversely, the final conclusion concerning lateral variations of the velocities
in the refractor cannot be drawn without a reliable estimate of the depth
structure.

4.2.1   The mean-minus-T method

This method has been used in Scandinavia since the middle of the 1950s for detailed determination of refractor velocities, particularly for assessing rock quality for underground constructions. The method is an application of the up-dip and down-dip concept for refractor velocity evaluations previously treated in Chapter 3. The difference between a general velocity determination technique and the mean-minus-T method is that the latter makes a more systematic use of the recorded time data. The name 'mean-minus-T' has been suggested by Parasnis (1984) to whom I am also indebted for the following concise mathematical explanation of the method.

If T_B and T_A are the arrival times at one and the same geophone from two impact points A and B and if

$$\Delta T = T_B - T_A$$

then obviously

$$\frac{\Delta T}{2} = \frac{T_B + T_A}{2} - T_A \tag{4.69}$$

The first term on the right-hand side of this equation is the mean of the arrival times from A and B while the second is the time T itself. This is the reason for calling the method the mean-minus-T method.

Now, if ΔT_n and ΔT_{n-1} are the time differences, as defined in Fig. 4.55, at two adjacent geophones separated by a distance S, the velocity $\bar{V}_2$ calculated from the relevant segment will be given by

$$
\begin{aligned}
\frac{1}{\bar{V}_2} &= \frac{1}{S}\left(\frac{\Delta T_n}{2} - \frac{\Delta T_{n-1}}{2} \right) \\
&= \frac{1}{S}\left(\frac{T_{B,n} - T_{B,n-1}}{2} + \frac{T_{A,n-1} - T_{A,n}}{2} \right) \\
&= \frac{1}{2}\left(\frac{1}{V_B} + \frac{1}{V_A} \right)
\end{aligned}
\tag{4.70}
$$

where V_B and V_A are simply the apparent forward and reverse velocities over the segment in question.

Equation (4.70) gives

$$\bar{V}_2 = \frac{2 V_B V_A}{V_B + V_A} \tag{4.71}$$

which is the harmonic mean of V_B and V_A and gives the true velocity if there is

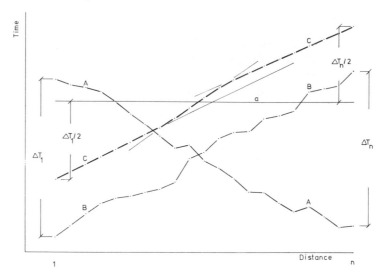

Fig. 4.55   The mean-minus-*T* method.

no dip. If the refractor is dipping between geophones n and $n-1$, the true velocity V_2 is given by

$$V_2 = \frac{2V_B V_A}{V_B + V_A} \cos\varphi \qquad (4.72)$$

The velocity determination technique has been described previously by Hagedoorn in 1959 (the plus-minus method) and by Hawkins in 1961 (the reciprocal method).

When the refractor as well as the ground surface are dipping, the relationship between the true velocity V_2 and the mean velocity $\bar{V}_2$ according to Equation (3.45) in Section 3.4 is

$$V_2 = \bar{V}_2 \cos(\varphi_2 - \varphi_1)/\cos\varphi_1 \qquad (4.73)$$

where φ_1 and φ_2 are the dips of the ground and refractor respectively. The signs of the angles must be consistent, i.e. minus for elevation and plus for depression angles, or vice versa.

The interpretation technique is illustrated in Figs 4.55 and 4.56. The time differences ΔT_1, etc. between the refractor velocity curves A and B are measured, divided by two and then plotted, with due regard to sign, from an arbitrary horizontal reference line (a). If the velocity calculation is made manually on a time–distance graph, the place of the reference line is so chosen that the resulting velocity determination curve C is not mixed up with other curves or interpretation lines. The points of curve C are to be connected by straight lines. An average line of 'best fit' should not be drawn through these points since, as will appear below, the segments between points may indicate

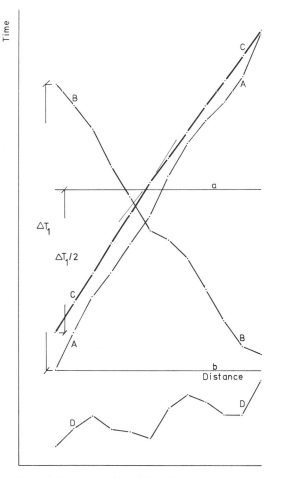

Fig. 4.56 Application of the mean-minus-*T* method.

significant velocity variations. For the analysis of the velocity lines it is convenient to start with the generally longer segments representing the higher velocities. When the slopes of these lines are settled, we can turn our attention to the segments with lower velocities. If the recorded and adjusted curves are marked in colour, the velocity interpretation lines can be drawn in black pencil. It is very likely that the interpreter changes his mind during the work and has to erase velocity evaluation lines.

The velocity evaluations are indicated in the figures by short extensions of the velocity segments. Intersections between these lines give approximately the velocity boundaries, or rather the locations on the ground surface of the wave front contacts. In Fig. 4.55 there are three different velocity segments in curve C. The higher velocities at the beginning and end of the profile are separated by a section with a lower velocity. The velocity line segment in the beginning of the profile is prolonged and we can observe that the two high-velocity segments are displaced. The passage of the waves through the lower velocity layer has caused a time delay. When evaluating the possible existence of a low-velocity zone in a refractor such time delays are just as interesting as the velocity lines themselves.

It is obvious that the recorded arrival times involved in the velocity determination must refer to the same refractor. A very common mistake is to mix up travel times from a solid bottom bedrock in one direction with those from an upper weathered and fractured rock layer with a lower velocity in the other direction. Special attention should be paid to low mean refractor velocities at the ends of a profile since they may be due to offset impact points being too close to the profile.

The technique can be used for any refractor as long as there are overlapping travel time curves from the direct and reverse recording. The velocities of the example in Fig. 4.56 lie within the range of soil material. It is difficult to get an acceptable value of the layer velocity from the recorded, irregular curves A and B and, furthermore, they do not give any indication of the velocity change detectable in the computed mean velocity in curve C. The 'half-time-differences' have been plotted from the reference line (a).

An ABC-curve can also be used to determine the mean refractor velocity. Curve D in Fig. 4.56 represents the sum of relative times from curves A and B. This curve and either curve A or B can be employed for the velocity determination. In the case shown, curve A has been corrected by taking half the time differences between curve D and the reference line (b) and adding these amounts to the corresponding arrival times in curve A. This procedure also leads to the adjusted times of curve C since the two approaches to the problem are basically the same.

4.2.2 Hales' method

Hales' method was originally designed for depth calculations but it has proved to be an excellent tool for detailed determinations of refractor velocities. Since the raypaths considered refer to a common point on the refractor, the method is particularly suited to solve problems connected with pronounced irregularities in the refractor surface or relatively great depths of the refractor and it is – at least theoretically – independent of depth. On the other hand, errors may be introduced because of varying near-surface conditions, thicknesses and velocities, between the different arrival locations of the rays on the ground. Too large variations in the upper layers may also preclude use of the method. Under favourable circumstances, i.e. with coverage in both directions of travel time curves from the upper layers, the ABC method can be used as a means of corrections. This will be demonstrated in Chapter 6 in connection with the analyses of some actual investigations. For deep refraction work we may apply conventional near-surface corrections to the raw travel times to have the impacts and the detectors adjusted to a horizontal or an inclined reference line. However, such corrections are generally not sufficiently accurate for detailed shallow refraction surveys.

The principles of Hales' method are discussed in Section 4.1.5. When the method is employed for refractor velocity determinations, the time loop is constructed in the usual way (Fig. 4.57) and the slope lines are drawn between the two limbs of the loop. The reciprocal slope of the lines equals $V_1 \sin i_{12} \cos \varphi_1 / \cos (\varphi_2 - \varphi_1)$ where φ_2 and φ_1 are the dips of the refractor and ground respectively. If the ground is horizontal but the refractor is inclined the term is reduced to $V_1 \sin i_{12} / \cos \varphi_2$. Omitting the dip angles, as recommended for depth determinations, may – under certain conditions – introduce errors. The next stage in the interpretation procedure is to establish the midpoints on the slope lines between the limbs of the time loop. A line connecting the midpoints yields the reciprocal slope of the calculated velocity, or velocities.

For a geometric proof of Hales' method applied to refractor velocity determinations, the structure and corresponding travel time curves of Fig.

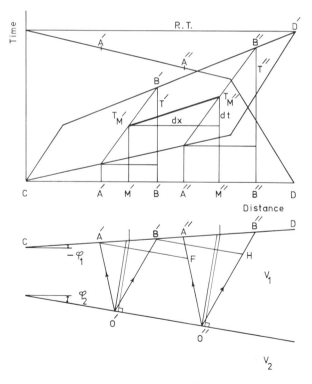

Fig. 4.57   Refractor velocity determination using Hales' method.

4.57 will be used. The time loop is closed between C and D'. The quantities T' and T'' are the travel times for the raypaths A'O'B' and A"O"B" respectively. M' and M" are the midpoints of the distances A'B' and A"B". The times $T_{M'}$ and $T_{M''}$ at the midpoints M' and M" are connected by a thick line. This centre line yields the calculated velocity V_2^-.

By definition, the velocity V_2^- is given by

$$\frac{dx}{dt} = V_2^- \tag{4.74}$$

We obtain dx and dt by expressing the various distances and times in terms of A'A" via the common refractor distance O'O". It is assumed that the waves are critically refracted into the V_1-layer and that the signs of the dip angles are consistent.

According to the figure

A'F = B'H = O'O"

$\angle A'O'B' = \angle A"O"B" = 2i_{12}$

$\angle A"A'F = \angle B"B'H = -\varphi_1 + \varphi_2$

$\angle O"A"A' = 90° + (i_{12} - \varphi_2 + \varphi_1)$

$\angle O"B"B' = 90° - (i_{12} + \varphi_2 - \varphi_1)$

$\angle A'FA" = 90° - i_{12}$

and

$\angle B'HB" = 90° + i_{12}$

$$dx = \left(A'A" + \frac{A"B"}{2} - \frac{A'B'}{2} \right) \cos\varphi_1$$

$$= \left(A'A" + \frac{B'B"}{2} - \frac{A'A"}{2} \right) \cos\varphi_1$$

$$= \frac{\cos\varphi_1}{2} (A'A" + B'B") \tag{4.75}$$

The terms have been multiplied by $\cos\varphi_1$ in order to get the horizontal distances.

The law of sines applied to the triangle B'HB" gives the following expression

$$B'B" = \frac{O'O" \cos i_{12}}{\cos(i_{12} + \varphi_2 - \varphi_1)} \tag{4.76}$$

From the triangle A′FA″, the distance O′O″ can be expressed in terms of A′A″ by the law of sines

$$O'O'' = \frac{A'A'' \cos(i_{12} - \varphi_2 + \varphi_1)}{\cos i_{12}} \tag{4.77}$$

so that

$$B'B'' = \frac{A'A'' \cos(i_{12} - \varphi_2 + \varphi_1)}{\cos(i_{12} + \varphi_2 - \varphi_1)} \tag{4.78}$$

Finally

$$dx = \frac{A'A'' \cos\varphi_1}{2} \left[1 + \frac{\cos(i_{12} - \varphi_2 + \varphi_1)}{\cos(i_{12} + \varphi_2 - \varphi_1)} \right] \tag{4.79}$$

The next step is to obtain dt in terms of A′A″

$$dt = T_{M''} - T_{M'}$$

where

$$T_{M'} = T'/2 + \text{R.T.} - T_{A'}$$

and

$$T_{M''} = T''/2 + \text{R.T.} - T_{A''}$$

R.T. being the reciprocal time.

$$dt = \frac{T'' - T'}{2} + T_{A'} - T_{A''}$$

$$= \frac{A''F + B''H}{2V_1} + \frac{O'O''}{V_2} - \frac{A''F}{V_1}$$

$$= \frac{O'O''}{V_2} + \frac{B''H - A''F}{2V_1} \tag{4.80}$$

The distance O′O″ is given above in terms of A′A″. The distances B″H and A″F can be expressed in A′A″ by applying the law of sines to the triangles B′HB″ and A′FA″ respectively.

$$B''H = \frac{O'O'' \sin(\varphi_2 - \varphi_1)}{\cos(i_{12} + \varphi_2 - \varphi_1)} \tag{4.81}$$

$$A''F = \frac{O'O'' \sin(\varphi_2 - \varphi_1)}{\cos(i_{12} - \varphi_2 + \varphi_1)} \tag{4.82}$$

The expression for dt is then

$$dt = \frac{A'A'' \cos(i_{12} - \varphi_2 + \varphi_1) \sin i_{12}}{\cos i_{12} V_1} + \frac{A'A'' \cos(i_{12} - \varphi_2 + \varphi_1) \sin(\varphi_2 - \varphi_1)}{\cos i_{12} 2V_1}$$

$$\times \left[\frac{1}{\cos(i_{12} + \varphi_2 - \varphi_1)} - \frac{1}{\cos(i_{12} - \varphi_2 + \varphi_1)} \right]$$

$$= \frac{A'A'' \cos(i_{12} - \varphi_2 + \varphi_1) \sin i_{12}}{\cos i_{12} V_1} + \frac{A'A'' \sin(\varphi_2 - \varphi_1) \sin i_{12} \sin(\varphi_2 - \varphi_1)}{\cos i_{12} \cos(i_{12} + \varphi_2 - \varphi_1) 2V_1}$$

$$= \frac{A'A'' \sin i_{12}}{V_1 \cos i_{12}} \left[\cos(i_{12} - \varphi_2 + \varphi_1) + \frac{\sin^2(\varphi_2 - \varphi_1)}{\cos(i_{12} + \varphi_2 - \varphi_1)} \right]$$

$$= \frac{A'A'' \sin i_{12}}{V_1 \cos i_{12}} \left[\frac{\cos^2 i_{12} \cos^2(\varphi_2 - \varphi_1) - \sin^2 i_{12} \sin^2(\varphi_2 - \varphi_1) + \sin^2(\varphi_2 - \varphi_1)}{\cos(i_{12} + \varphi_2 - \varphi_1)} \right]$$

$$= \frac{A'A'' \sin i_{12} \cos^2 i_{12} [\cos^2(\varphi_2 - \varphi_1) + \sin^2(\varphi_2 - \varphi_1)]}{V_1 \cos i_{12} (i_{12} + \varphi_2 - \varphi_1)}$$

$$= \frac{A'A'' \sin i_{12} \cos i_{12}}{V_1 \cos(i_{12} + \varphi_2 - \varphi_1)} \tag{4.83}$$

Since dx as well as dt is now expressed in terms of $A'A''$, we can write

$$V_2^- = \frac{dx}{dt} = \left\{ \frac{A'A'' \cos\varphi_1 [\cos(i_{12} + \varphi_2 - \varphi_1) + \cos(i_{12} - \varphi_2 + \varphi_1)]}{2\cos(i_{12} + \varphi_2 - \varphi_1)} \right\} \Big/$$

$$\left\{ \frac{A'A'' \sin i_{12} \cos i_{12}}{V_1 \cos(i_{12} + \varphi_2 - \varphi_1)} \right\}$$

$$= \frac{A'A'' \cos\varphi_1 [\cos(i_{12} + \varphi_2 - \varphi_1) + \cos(i_{12} - \varphi_2 + \varphi_1)] V_1 \cos(i_{12} + \varphi_2 - \varphi_1)}{2\cos(i_{12} + \varphi_2 - \varphi_1) A'A'' \sin i_{12} \cos i_{12}}$$

$$= \frac{V_1 \cos\varphi_1 [\cos(i_{12} + \varphi_2 - \varphi_1) + \cos(i_{12} - \varphi_2 + \varphi_1)]}{2\sin i_{12} \cos i_{12}}$$

$$= \frac{V_1 \cos\varphi_1 \cos(\varphi_2 - \varphi_1)}{\sin i_{12}} \tag{4.84}$$

If $V_1/\sin i_{12}$ is replaced by V_2, the true refractor velocity, we have the relation

$$V_2^- = V_2 \cos\varphi_1 \cos(\varphi_2 - \varphi_1) \tag{4.85}$$

In the case of dipping interfaces, V_2^- is always lower than the true velocity. The refractor velocity V_2^- computed by Hales' method is not the same as the harmonic mean $\bar{V}_2$ obtained by the mean-minus-T method. Dipping interfaces can cause the latter velocity to be too high or too low in relation to the true velocity V_2.

Because of the dipping interfaces, the midpoints M′ and M″ do not coincide with the intersections between the ground surface and the bisectors to the angles A′O′B′ and A″O″B″. Therefore, the midpoints projected normally on to the refractor surface are somewhat displaced in relation to the points O′ and O″, as can be seen in Fig. 4.57.

For multilayer cases, a refractor velocity analysis using Hales' method has to be preceded by the determination of velocities and thicknesses of the various overburden layers to enable the localization on the ground surface of the rays from the common point on the refractor to be made. The general picture of the refractor velocities obtained by the mean-minus-T method also forms a basis for the detailed interpretation by Hales' method.

For a simple two-layer case with horizontal interfaces, the inclination of the slope lines depends on the distance $2h_1 \tan i_{12}$ between the emergence points on the ground and the total travel time $2h_1/V_1 \cos i_{12}$ for the slant raypaths. Thus, the inclination is $2h_1 \tan i_{12} V_1 \cos i_{12}/2h_1 = V_1 \sin i_{12}$. The distance between the points on the ground for multilayer cases is given by the term

$$2 \sum_{1}^{n-1} d_v \tan i_{vn} \tag{4.86}$$

and the corresponding travel time by

$$2 \sum_{1}^{n-1} d_v/V_v \cos i_{vn} \tag{4.87}$$

In the case of Fig. 4.58, $n = 4$ and V_v and d_v are the velocities and thicknesses of the layers from 1 to 3. There are, however, simpler ways to calculate the inclination of slope lines and still have an acceptable solution for the velocity distribution.

The multilayer case can be transformed into a two-layer case by using the equation

$$\sum_{1}^{n-1} V_v d_v = V_x \sum_{1}^{n-1} d_v \tag{4.88}$$

where V_v and d_v are the velocities and thicknesses of the layers. The resulting fictitious, weighted velocity V_x is then used to compute the inclination of the slope lines according to the term

$$V_x \sin i_{xn} \tag{4.89}$$

This technique has been used to determine the velocities in the structures in Figs 4.58–4.60. Another way to reduce a multilayer case to a two-layer case is to calculate a fictitious velocity-term yielding the total depth of the refractor in question.

In the four-layer structure of Fig. 4.58 the interfaces are assumed to be horizontal and the velocities to be constant except for a velocity change in the bottom refractor, $V_{4'} < V_4$. The entire traverse is covered from both directions by arrival times emanating from the fourth layer, curves 1 and 2.

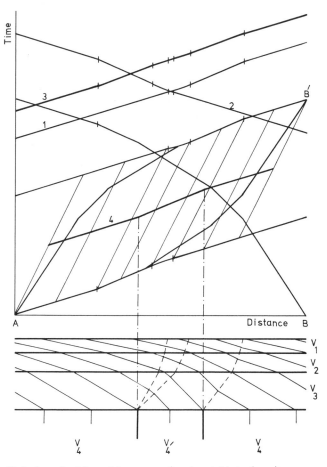

Fig. 4.58 Hales' method for a 4-layer case (horizontal interfaces).

The wave fronts of the forward recording from offset impact point 1 are also drawn in Fig. 4.58. The wave front contact and the diffraction zone are marked by dashed lines. Because of the symmetry of the structure, impact 2 presents the mirror image of the wave front pattern of 1.

The problem to be solved is to determine the velocity distribution in the bottom layer, i.e. to locate with an acceptable accuracy the boundaries for the $V_{4'}$ zone.

The time loop in the figure is closed between A and B'. In order to complete the loop, curves 1 and 2 have been tied to the respective bottom refractor velocity segments from A and B. The inclination of the slope lines to be drawn between the two limbs of the loop is based on a calculated, weighted velocity term V_x according to Equation (4.88). When the velocities V_x and V_4 are used to calculate $\sin i_{xn}$, the procedure leads to an overestimate of the inclination of the slope lines in the parts of the structure where the velocity of the fourth layer is V_4 but to an underestimate above the $V_{4'}$ zone. This does not seem to affect the results appreciably. The midpoints of the slope lines are connected in the usual manner to a centre line (curve 4), the velocity line segments of which yield the velocities and their boundaries. This simplified interpretation technique gives an acceptable result. The intersections between the inverse velocity lines in curve 4 coincide with the velocity boundaries in the bottom refractor as shown by the vertical dash-dotted lines.

Curve 3 at the top of the graph in Fig. 4.58 indicates the velocities when the mean-minus-T method is used for the velocity calculation. This interpretation results in a too wide zone – between the wave front contacts on the ground surface – and, as a consequence, the calculated velocity of the $V_{4'}$ zone is overestimated. The velocity analysis can be improved by tracing the wave front contacts back to the bottom refractor or by an employment of a displacement technique to be discussed later on.

The simplified velocity interpretation yields acceptable results also in the case of dipping interfaces. In Fig. 4.59 curves 1 and 2 refer to offset impact points giving the apparent velocities from the bottom refractor, the third layer. It is assumed that $V_{3'}$ is less than V_3. The time loop runs between A and B'. For the completion of the time loop, curves 1 and 2 have been used. Note that to the right of point A, the velocity line from the overturned reverse recording almost coincides with the zero time line. The refractor velocity (mean-minus-T) is given by curve 3 while curve 4 refers to Hales' method. The inclination of the slope lines is obtained by the same simplified procedure as for horizontal layers, i.e. using Equations (4.88) and (4.89), without considering the possible influence of the layer dips. The boundaries between the velocity sections (dash-dotted lines) have to be projected perpendicularly to the bottom refractor surface. However, a perfect match is not possible to obtain since, as mentioned for the simple two-layer case in Fig. 4.57, the midpoints do not lie on the intersection between the angle bisectors and the ground surface.

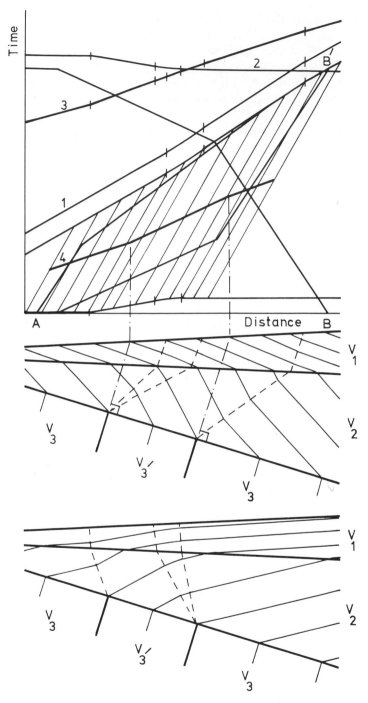

Fig. 4.59 Hales' method for a 3-layer case (dipping interfaces).

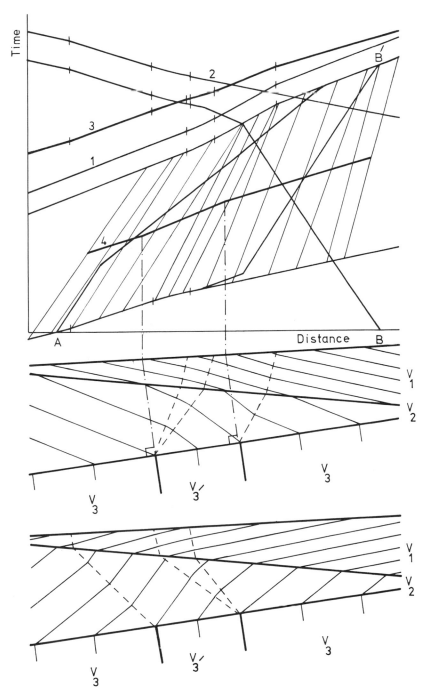

Fig. 4.60 Hales' method for a 3-layer case (dipping interfaces).

Figure 4.60 shows a case where the layers dip in different directions. The inclination of the slope lines has to be varied along the traverse. Since there is a continuous change, average inclination values are employed for shorter parts of the section. As in previous examples, curves 3 and 4 refer to mean-minus-T and Hales' method respectively. The inclination of the slope lines is based on Equations (4.88) and (4.89).

4.2.3   Comparison between refractor velocity determination methods

The two refractor velocity determination methods, mean-minus-T and Hales', give partly different results. The mean velocities obtained differ and they have to be adjusted for dipping interfaces in different ways. In selecting the method to be used we must take into consideration the type of project, the geology and the limitations of the methods. It is advisable to commence with the mean-minus-T method to get a general picture of the refractor velocities which, moreover, is needed for the depth determinations.

The effects of the different raypath systems used are outlined in Figs 4.61–4.62. The upper cross-section (a) in Fig. 4.61 refers to the mean-minus-T method. Because of the slant raypaths through the upper medium, the arrival times of curves 1 and 2 over the distance Δx on the ground emanate from two separate sections of the refractor. With decreasing depth and/or incidence angle, the sections on the refractor partly overlap. The velocity analysis for the distance Δx yields an average value for two different parts of the refractor. As long as the conditions in the refractor are rather uniform, the mean-minus-T method gives acceptable results for the velocity distribution. On the other hand, relatively great depths of the refractor and/or dip changes in the refractor may obscure minor velocity details or produce false velocity segments. An advantage with the mean-minus-T method is that the effects of near-surface irregularities, in velocities or thicknesses, are eliminated or at least minimized since the rays involved emerge at the same points on the ground surface.

In the lower part of Fig. 4.61, the raypaths considered in using Hales' method are given. Curves 1 and 2 are the limbs in the time loop. Since the interfaces are horizontal (parallel), the curves are parallel. The location of the arrival times of waves coming from the refractor section Δx are drawn in the travel time curves as thick lines. Curve 3 represents the centre line. The midpoints on the slope lines are vertically above the endpoints of the refractor section Δx as indicated by the dash-dotted lines. Contrary to the mean-minus-T method, Hales' method yields here the velocity in, and the position of, a particular refractor section and not an average value for a longer part of the refractor. It must be remembered, however, that an application of Hales' method may be hampered or even made impossible by large, lateral variations in the uppermost layers since the rays involved emerge at different points on the ground surface.

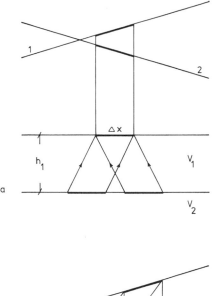

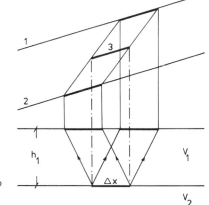

Fig. 4.61   Comparison between mean-minus-T and Hales' method.

The raypath systems for dipping refractor are studied in Fig. 4.62. In this case the centre line (3) when projected perpendicularly on to the refractor is somewhat displaced in relation to the section Δx. The reason is that the midpoints on the slope lines do not correspond to the points at which the bisectors of the angles between the rays from the refractor meet the ground surface. The resulting error increases when the dip angles for the ground and the refractor are in different quadrants and decreases when they lie in the same quadrant.

In Fig. 4.63 the problem of slant raypaths is regarded from another view-point. Both cross-sections in the figure refer to the mean-minus-T method. In Figs 4.61 and 4.62 the studies were based on the common registrations on the

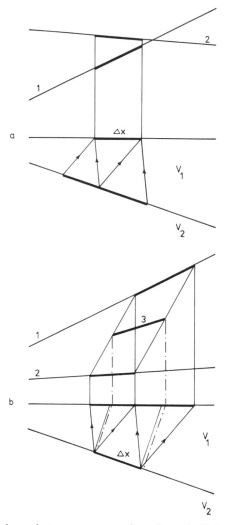

Fig. 4.62 Comparison between mean-minus-T and Hales' method (dipping refractor).

ground. Figure 4.63 shows how the waves from a particular refractor section Δx are displayed in the arrival times in the direct and the reverse recording. If curve 2 in (a) is displaced $2h_1 \tan i_{12}$ to the right, as indicated by the arrows, the arrival times in (1) and (2) will have the same horizontal position. An application of the mean-minus-T method now yields the true velocity for the section Δx. The calculated mean refractor velocity curve has then to be displaced $h_1 \tan i_{12}$ to the left, half the displacement distance, as shown on the graph by an arrow. The problem cannot be solved in such a simple way for the case in (b). Shallow features are displaced less than the deeper ones in the

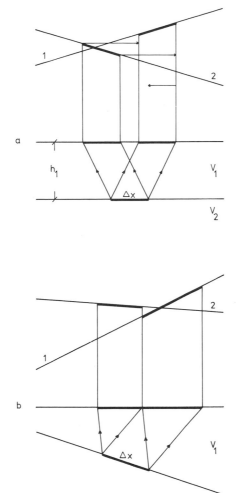

Fig. 4.63   Slant raypaths from another viewpoint (mean-minus-*T* method).

travel time curves, and, as can be seen on the graph, the velocity segments in the direct and reverse recording are not equal. A displacement by an average value will improve the velocity interpretations but errors are due to remain. In Hales' method depth variations are automatically taken account of.

The widening of refracted and diffracted velocity segments with increasing thickness of the overburden and the errors in the corresponding computed mean velocities are illustrated in Fig. 4.64. It is assumed that the overburden velocity V_1 is constant and that $V_{2'}$ is less than V_2. The dashed horizontal lines on the cross-section, marked 1–4, correspond to varying overburden levels.

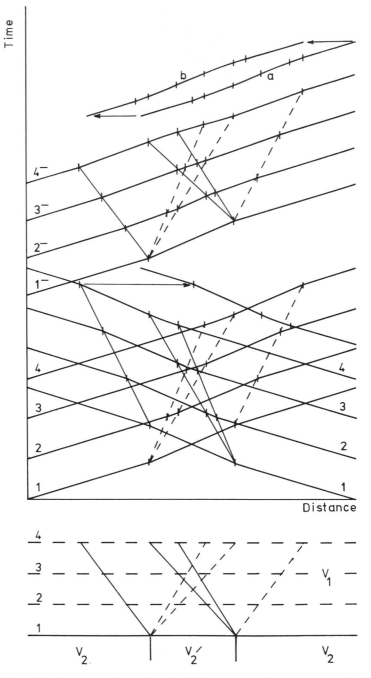

Fig. 4.64 Widening of refracted and diffracted velocity segments (horizontal refractor).

At level 1 there is no overburden at all. The corresponding recorded travel time curves are also marked 1–4. In the direct and reverse recording, the widening at the various levels of the diffraction zones and the wave front contacts are given by dashed and solid lines respectively. The curves $1^- – 4^-$ represent the mean refractor velocities when the mean-minus-T method is employed. In order to keep the figure to a reasonable size the angles of incidence are exaggerated.

Curve 1^- gives the true velocities in the refractor and the position of the velocity boundaries. For the second level, curve 2^-, only in the central part of the low-velocity zone is the true velocity $V_{2'}$ obtained. This velocity segment is surrounded by two segments composed of an average of the velocity $V_{2'}$ and the velocities in the diffraction zones. At the beginning and end of the zone in curve 2^-, the mean velocity lines yield an average of V_2 and $V_{2'}$. The computed mean velocity in curve 3^- consists of average values of V_2, $V_{2'}$ and velocities from the diffractions. The velocity $V_{2'}$ is not present in the calculated mean velocity. At the next level, 4, the diffraction zones in the direct and the reverse recording have changed places. In the central part of the low velocity in the refractor there is a short section where the computed mean velocity is not $V_{2'}$ but V_2. With still greater depth to the refractor the interpretation will lead to a section with velocity V_2 above the zone of actual velocity $V_{2'}$ and on both sides of this V_2 section two minor velocity zones in areas where the true velocity is V_2. When the depth is large in relation to the width of a low-velocity zone, indications of the lower velocity may be overlooked and regarded as an insignificant undulation in the computed mean velocity curve.

Since the interfaces were assumed to be parallel, evaluation of the velocities and their boundaries can be improved by displacing one curve. On the time plot in Fig. 4.64, curve 4 from the reverse recording has been displaced $2h_1 \tan i_{12}$ in the direction of the arrow. A renewed velocity calculation using the displaced curve and curve 4 from the forward recording yields curve (a) at the top of the graph. This velocity curve must then be removed to position (b), half the displacement distance, to get the horizontal position of the velocity boundaries. Strictly speaking, the displacement ought to be $h_1 \tan i_{12} + h_1 \tan (i_{12} + i_{12'})$ and not $2h_1 \tan i_{12}$ but, in practice, the latter term will suffice.

A refractor velocity analysis based on Hales' method gives an acceptable result for all levels in Fig. 4.64 since that technique (theoretically) is independent of the depth to the refractor.

When the interfaces are parallel or almost parallel, the intersections between the velocity segments obtained by Hales' method are in agreement with the velocity boundaries in the refractor. For dipping, non-parallel layers only a close approximation is to be expected as can be seen from Fig. 4.65. The discrepancy between the calculated velocity boundaries and the real ones in the refractor decreases when the dip angles approach each other. If, for some reason, Hales' method or displacement technique cannot be used to calculate the velocities and their boundaries, a technique of tracing the wave front contacts back to the emergence points on the refractor will improve results obtained by the mean-minus-*T* method. Curve 3 in the figure is calculated using the latter method. The computed zone is too wide in relation to the real one in the refractor and the velocity value obtained is overestimated. Since the velocity line intersections in curve 3 coincide with

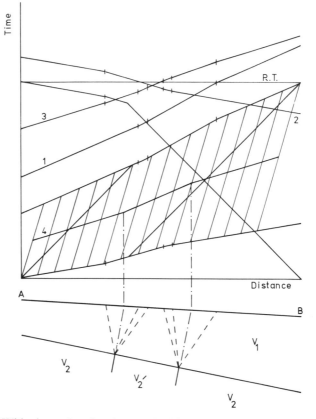

Fig. 4.65 Widening of refraction and diffraction velocity segments (dipping refractor).

the wave front contacts on the ground surface, this information can be used to plot the corresponding wave front locations on the refractor. The angles which the up-dip and down-dip wave front contacts make with a vertical line are $(i_{12} + i_{12'})/2 - \varphi_2$ and $(i_{12} + i_{12'})/2 + \varphi_2$ respectively. When the depths are calculated and the structure is plotted, these angles can be used to trace the location of the velocity boundaries in the refractor from the indications of the wave front contacts on the ground. The velocity value for $V_{2'}$ can then be improved by making use of the better estimate of the width of the $V_{2'}$ zone.

As has been demonstrated previously, the mean-minus-T method may give false velocity segments in the case of dip changes in the refractor. Referring to Fig. 4.42 we see that there is a low-velocity segment in curve 1 between the wave front contacts on the ground despite the fact that the refractor velocity is constant. Hales' method employed for the determination of the velocity would indicate that there is no velocity change in the refractor provided that

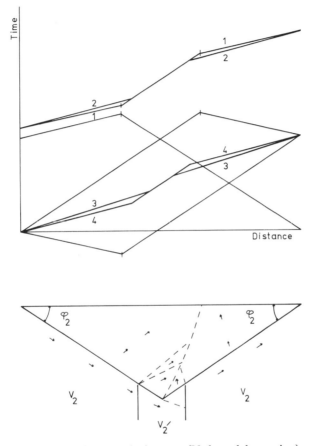

Fig. 4.66 Displacement of a low-velocity zone (V-shaped depression).

the dip angle φ_2 was taken into account when drawing the slope lines, otherwise a smaller velocity step will remain in the curve.

In Fig. 4.66 there is a velocity change in the refractor, $V_{2'} < V_2$. Other pertinent data are the same as those of Fig. 4.42. The mean velocity lines obtained by the mean-minus-T method are plotted as 1 and 2 in the graph. The velocity segments in curves 3 and 4 are obtained using Hales' method. In this case both methods yield a low-velocity segment above the $V_{2'}$ zone.

Owing to the dip of the refractor, curve 1 gives a velocity higher than V_2. In curve 2 the calculated velocity has been corrected for the influence of the dip, i.e. the velocity value from curve 1 has been multiplied by $\cos \varphi_2$. On the other hand, the velocity in curve 3 is underestimated and has to be divided by the cosine of the dip angle. The corrected velocity value is plotted as curve 4.

Neither of the methods employed gives the correct location of the low-velocity zone nor the velocity in it. However, the main problem, namely to ascertain the existence of the velocity change in the refractor, has been solved.

For an asymmetric depression in the refractor, the interpretations displace a low-velocity zone towards the flank with the smaller dip angle. The midpoints of the low-velocity segments in Fig. 4.67 lie to the left of the centre of the zone in the refractor. The same situation was observed in Fig. 4.50 where the midpoint of the false low-velocity segment and maximum depth obtained are displaced to the left of the dip change in the refractor. As in Fig. 4.66 curves 1 and 3 give velocities uncorrected for dip, while curves 2 and 4 are velocity curves adjusted by means of the cosine of the dip angles.

The shape of a depression, maximum depth, velocity boundaries and true velocities are difficult to determine. However, by an appropriate combination of various interpretation techniques it is possible to come closer to reality.

In the case of large scale faulting, the velocity analysis is greatly facilitated if both the mean-minus-T method and Hales' method can be applied to the problem. Figs 4.68 and 4.69 show the same structure except for the low-velocity zone on the downthrown side of the refractor in Fig. 4.69. The velocity picture, curves 1, when the mean-minus-T method is used is rather similar in both figures. The calculated velocities (inverse slopes) increase towards the edge of the fault to become infinite immediately to the right of the edge. The combination of the arrival times from the edge and those from the downthrown side in the reverse recording gives rise to a low-velocity segment, apparent in the case of Fig. 4.68 and partly real in that of Fig. 4.69.

The line connecting the midpoints of the slope lines in Fig. 4.68, curve 2, has a minor step caused by a computed velocity above the fault that is too high, while in Fig. 4.69 curve 2 indicates a more pronounced low-velocity segment. For the determination of the inclination of the slope lines V_1 and V_2 have been used. The dashed line in the time plot in Fig. 4.69 refers to a recalculation of the velocity where the slope lines, dash-dotted lines, are based on V_1 and an estimate of $V_{2'}$.

If the seismic line is sufficiently extended to the left on the upthrown side of the fault, the existence of a possible low-velocity zone can be checked by an extension of the velocity segments on the left side of the plots towards the right. In Fig. 4.69, contrary to the case in Fig. 4.68, there is a considerable time delay between these lines and the calculated velocity curves at the end of the profile.

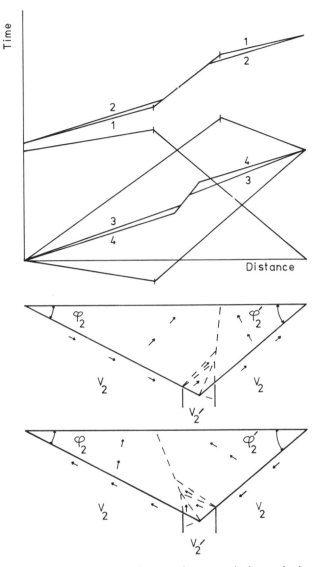

Fig. 4.67 Displacement of a low-velocity zone (asymmetric depression).

Pitfalls associated with the mean-minus-T method can be avoided by an application of Hales' method. The latter method has proved particularly useful for the determination of the presence or otherwise of shear zones and to pinpoint velocity boundaries in the refractor. On the other hand, velocity boundaries obtained by the mean-minus-T method indicate sections where the depth determinations are uncertain.

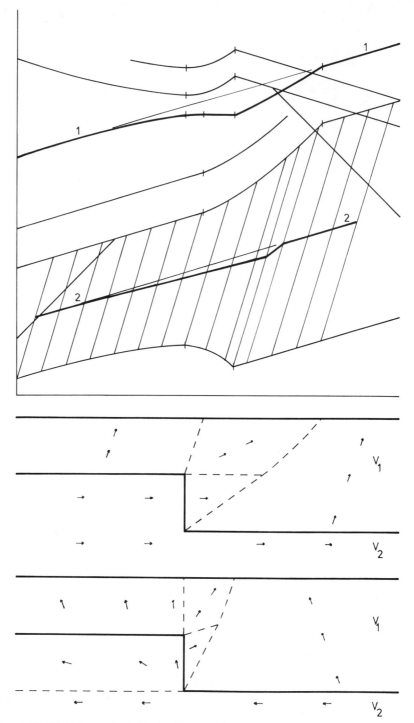

Fig. 4.68 Velocity analysis (faulted structure).

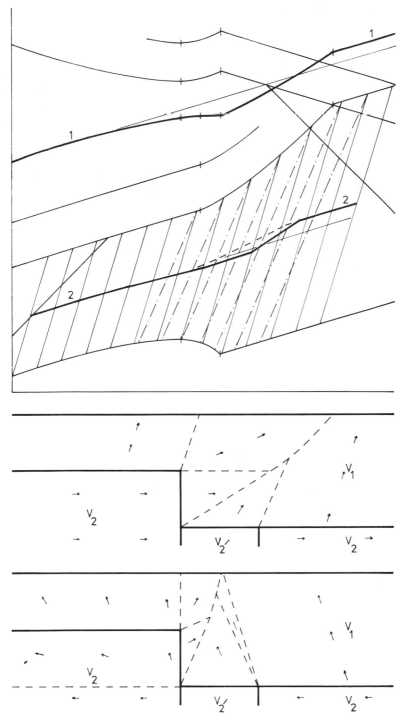

Fig. 4.69 Velocity analysis (faulted structure).

4.2.4   Non-critical refraction

In the literature on refraction seismics it is generally assumed that the waves are critically refracted. In reality, we have to reckon with a considerable amount of non-critically refracted waves, besides the diffracted waves. Even rather moderate variations in the refractor surface lead to travel paths of least time through the structures and not along the interfaces. In Fig. 4.24 the rays in the direct recording make an angle with the refractor surface between M and N and to the right of Q. In the reverse recording the waves are non-critically refracted between Q and N and to the left of M. A mean refractor velocity determination by the mean-minus-T method is affected in this case along the entire traverse by non-critical refraction and diffraction zones. The erroneous velocities caused by non-critical refraction are always higher than the true velocities and the incidence angles are smaller than the critical angles.

If the incoming rays make an angle β with a horizontal interface, the refracted waves in the overlying medium make an angle i_{12}^- with the interface; i_{12}^- is less than i_{12}. Then according to Snell's law

$$\sin i_{12}^- = \frac{V_1 \cos \beta}{V_2} = \sin i_{12} \cos \beta \tag{4.90}$$

In order to get the true velocity V_2, the recorded apparent velocity has to be multiplied by $\cos \beta$.

In the examples presented previously, we can observe that often the rays from one side make an angle with the refractor surface, when the waves travel along the refractor in the other direction. The resulting calculated velocity will be

$$\bar{V}_2 = \frac{2V_1}{\sin i_{12} + \sin i_{12}^-} = \frac{2V_2}{1 + \cos \beta} \tag{4.91}$$

As a numerical example, assume that $V_2 = 5500$ m/s. For angles $\beta = 10°, 20°$ and $30°$ the calculated mean refractor velocities are 5542, 5670 and 5895 m/s respectively.

For dipping refractors the velocity relation is more complicated. In Fig. 4.70 the rays in the forward recording impinge obliquely on the dipping surface of the V_2-layer. It is assumed that in the other direction the waves travel along the surface.

According to the figure

$$AD = x \cos (i_{12}^- + \varphi_2)/\cos i_{12}^-$$

$$CD = x \cos (i_{12}^- + \varphi_2) \cos \beta / \cos i_{12}^-$$

$$DB = x \sin \varphi_2 / \cos i_{12}^-$$

$$R = 90° - \beta$$

$$\angle ABD = 90° - i_{12}^- - \varphi_2$$

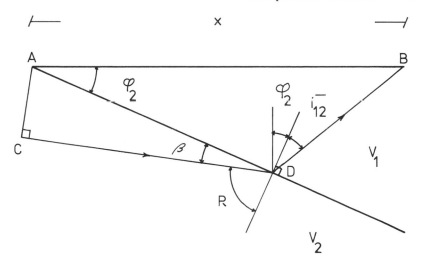

Fig. 4.70 Non-critical refraction (dipping refractor).

The travel time from C to B is

$$t = \frac{x\cos(i_{12}^- + \varphi_2)\cos\beta}{V_2\cos i_{12}^-} + \frac{x\sin\varphi_2}{V_1\cos i_{12}^-} \qquad (4.92)$$

where

$$V_2 = \frac{V_1\cos\beta}{\sin i_{12}^-}$$

The slope of the down-dip velocity line will be

$$\frac{dt}{dx} = \frac{\cos(i_{12}^- + \varphi_2)\sin i_{12}^-}{V_1\cos i_{12}^-} + \frac{\sin\varphi_2}{V_1\cos i_{12}^-} = \frac{\sin(i_{12}^- + \varphi_2)}{V_1} \qquad (4.93)$$

In the reverse recording the slope of the velocity line is $\sin(i_{12} - \varphi_2)/V_1$ as the waves are assumed to be critically refracted into the upper medium. In order to get the mean velocity $\bar{V}_2$ the slopes have to be added and divided by two.

Thus

$$\frac{1}{\bar{V}_2} = \frac{1}{2}\left[\frac{\sin(i_{12}^- + \varphi_2)}{V_1} + \frac{\sin(i_{12}^- - \varphi_2)}{V_1}\right]$$

$$= \frac{\sin\dfrac{i_{12}^- + i_{12}^-}{2}\cos\left(\varphi_2 - \dfrac{i_{12}^- - i_{12}^-}{2}\right)}{V_1}$$

Replacing V_1 by $V_2 \sin i_{12}$, we obtain

$$\bar{V}_2 = V_2 \sin i_{12} \bigg/ \left[\sin \frac{i_{12} + \bar{i}_{12}}{2} \cos \left(\varphi_2 - \frac{i_{12} - \bar{i}_{12}}{2} \right) \right] \tag{4.94}$$

If we assume that the true velocities are 1500 m/s and 5000 m/s, the dip angle $\varphi_2 = 20°$ and $\beta = 20°$, then the calculated mean velocity will be 5475 m/s.

4.2.5 Influence of angle between strike of structure and seismic profile

In the previous sections it has been assumed that the seismic measurements have been carried out perpendicularly to the strike of the structure and in the azimuth of maximum dip, because it is advantageous in developing the basic theory to treat the various factors one by one instead of involving them all together. In reality, however, a number of different factors influence the recorded arrival times. The chance of success in solving particular geological problems can be increased by measuring profiles in different directions or, in other words, by acquiring more information from the subsurface. A complementary investigation is necessary if the first measured profile happens to make a non-right angle with the strike of the structure, since completely false results may be obtained when the angle is small or when the seismic line is parallel to the structure.

For simplicity it is assumed in the examples studied below that the interfaces are horizontal and that the boundaries between velocity sections in the refractor are vertical while the seismic profiles in plan make acute angles with, or are parallel to, the structure.

The wave front pattern and velocities obtained when waves impinge obliquely on velocity boundaries in the refractor will be discussed for the structures of Figs 4.71 and 4.72. In Fig. 4.71 the waves pass through a low-velocity zone, a case of great interest since velocities are often used to assess rock quality for engineering purposes and water prospecting. Wave propagation from a lower to a higher velocity section is studied in Fig. 4.72. The figures show the situation in plan.

The measuring line AA in Fig. 4.71 makes the angle β with the boundaries of the low-velocity section, $V_{2'} < V_2$. Since the raypaths make a non-right angle with the boundary between the high and low-velocity sections they are refracted in the $V_{2'}$ zone towards the normal to the boundary. The wave direction is shown by the arrows. The incidence angle $i_{22'}$ is $90° - \beta$. The refraction angle R is given by the equation

$$\sin R = \frac{V_{2'} \sin \bar{i}_{22'}}{V_2}$$

The rays in the low-velocity zone make the angle $\bar{i}_{22'} - R$ with the measuring line and an apparent velocity $V_{2'app}$ will be recorded, $V_{2'app} > V_{2'}$. The apparent velocity is obtained by the equation

$$V_{2'app} = \frac{V_{2'}}{\cos{(\bar{i}_{22'} - R)}} \tag{4.95}$$

With decreasing angle between the structure and the line of measurements, the error in velocity will be significant, particularly if there is a large contrast between velocities.

As a numerical example, suppose that $V_2 = 5500$ m/s, $V_{2'} = 4000$ m/s and the angle of incidence $\bar{i}_{22'}$ is 75°. The recorded velocity $V_{2'app}$ will then be about 4640 m/s. In hard rocks a shear zone with a velocity of 4000 m/s, compared with the velocity 5500 m/s for the compact rock of the area, is to be regarded as critical for excavation works. The apparent velocity (4640 m/s) does not indicate the real rock conditions and unexpected, great difficulties may be encountered during the construction work if no further inspection by additional profiles or drilling is made.

Since the velocity boundaries in Fig. 4.71 are assumed to be parallel, the rays, when they leave the low-velocity zone and enter the V_2-layer in the right-hand part of the figure, are parallel to the seismic line AA. The waves recorded in the seismic line to the right of the $V_{2'}$ zone follow the path BCDA.

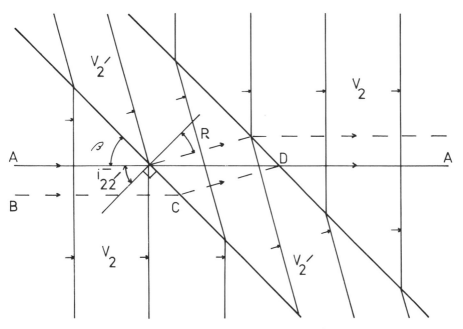

Fig. 4.71 Effect of strike angle (waves passing from high-velocity zone to low-velocity zone).

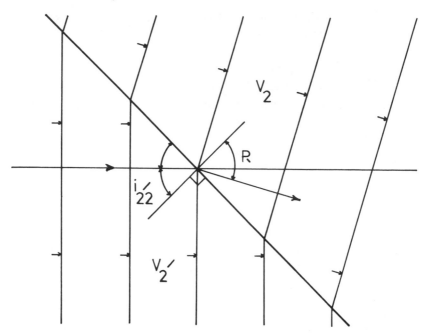

Fig. 4.72 Effect of strike angle (waves passing from low-velocity zone to high-velocity zõne).

Wave propagation from a low- to a high-velocity layer is considered in Fig. 4.72. In this case the rays are refracted away from the normal to the boundary in the section with the higher velocity V_2. The refraction angle R and the apparent velocity V_{2app} are given by the relations

$$\sin R = \frac{V_2 \sin i_{2'2}^-}{V_{2'}}$$

and

$$V_{2app} = \frac{V_2}{\cos(R - i_{2'2}^-)} \tag{4.96}$$

As an example, assume that $V_2 = 5000$ m/s, $V_{2'} = 4000$ m/s and $i_{2'2}^- = 45°$. The recorded velocity V_{2app} will then be about 5230 m/s.

In Fig. 4.73 AB represents the line of measurements, A and B being the impact points. The wave front propagation is shown in the horizontal plane with the reverse and direct recordings in (a) and (b) respectively. In order to simplify the model study it is assumed that there are no overburden layers. The seismic line crosses the velocity boundary ($V_{2'} < V_2$) at D.

In the direct recording (b), the waves in the $V_{2'}$-layer are recorded between

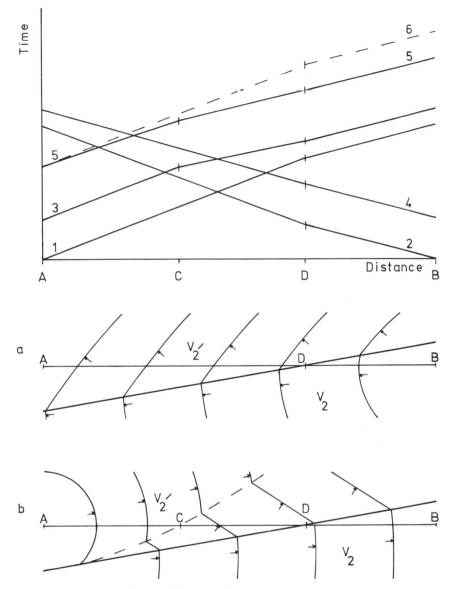

Fig: 4.73 Effect of strike angle (outcropping refractors).

A and C. Point C is the wave front contact (dashed line) in the seismic line. From C to D the waves critically refracted from the V_2-layer overtake. Between D and B the waves in the V_2-layer are recorded. In the reverse recording (a) the velocity V_2 is obtained from B to D. From the latter point the waves refracted from the V_2-layer are the first to reach the measuring line.

Curves 1 and 2 on the time–distance graph indicate the true velocities $V_{2'}$ and V_2, while curves 3 and 4 refer to the actual, recorded travel times. The model is constructed using the velocity 4000 m/s for the $V_{2'}$-layer and 6000 m/s for the V_2-layer. In curve 3, the direct recording, the true velocity 4000 m/s is recorded between A and C. Over the distance CD, where the first arrivals correspond to waves refracted from the V_2-layer, the apparent velocity lies around 7500 m/s. At the end of the line, D to B, the recorded velocity approaches 6000 m/s. The velocity is actually somewhat higher than 6000 m/s as the rays make an angle with the measuring line. The angle decreases with increasing distance from D. In the reverse recording the true velocity 6000 m/s is obtained between B and D, while in the $V_{2'}$-layer from D to A the waves refracted from the V_2-layer give the velocity 5200 m/s.

The mean refractor velocities, based on the recorded travel time curves 3 and 4, are given by curve 5, which is constructed using the mean-minus-T method. As can be seen, curve 5 has three velocity segments. The only acceptable velocity determination is to be found at the end of the seismic line, between D and B, where the mean velocity is near 6000 m/s. In the beginning of the profile, the distance AC, the calculated velocity is 4600 m/s, thus 600 m/s higher than the true one. However, the most serious misinterpretation is between C and D where the mean velocity is about 6200 m/s, even higher than in the V_2-layer. To facilitate a comparison, the curve of true velocities, namely curve 6, has been plotted adjacent to the mean velocity curve 5.

When the directions of the structure and the measurements are parallel or almost parallel, not only do the higher velocities tend to obscure the lower ones but also the depth analysis may be faulty. The travel time curves of Fig. 4.74 indicate without doubt that they correspond to a three-layer case. In the travel time curves from A and B there are three discrete velocity segments and the third velocity lines are parallel to the travel time curves from the offset impact points 1 and 2. The obvious, seemingly faultless interpretation is given in the cross-section.

However, an additional measuring line (Fig. 4.75) perpendicular to the line AB in Fig. 4.74 yields quite another geological picture, namely a two-layer case with a velocity change ($V_{2'} < V_2$) in the refractor below profile AB. The profiles intersect each other at point C.

The explanation of the recorded travel time curves of Fig. 4.74 is that the arrival times of the 'second layer' refer to the $V_{2'}$-layer and that the third velocity segments emanate from waves travelling with the velocity V_2 by the longer but faster paths in the refractor sections surrounding the low-velocity zone. In line AB a velocity change in the refractor has been transformed in the interpretation into a horizontal layer, depths have been misinterpreted and the low-velocity section remains undetected because of insufficient measuring data.

Since the arrival times recorded parallel to and above a low-velocity zone depend on the depth of the refractor, the width of the zone and the velocity

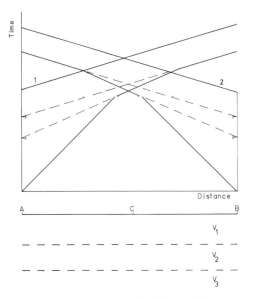

Fig. 4.74   An apparently three-layer case (cf. Fig. 4.75 for true interpretation).

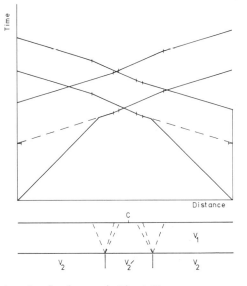

Fig. 4.75   True situation for the case in Fig. 4.74.

contrasts, another case can be postulated where there are no registrations from the lower refractor velocity and only V_1 and V_2 appear in the travel time curves. A depth analysis based on these data will not give the vertical depth to the refractor but the distances from the seismic line to the surrounding high-velocity sections.

5

Instrumentation, field work and interpretation procedure

There are three components in a seismic survey, namely instrumentation, field work and interpretation, which must be well tuned to each other if reliable results are to be obtained. The time data provided by the equipment must have sufficient resolution, the field operations must yield adequate information and the interpretation work must do justice to the field data obtained. A fourth factor affecting the reliability of the results is the positioning of the seismic lines with regard to the geology and the possibilities and limitations of the method.

5.1 EQUIPMENT

Only the main principles of seismic instruments will be described here. For details of the various kinds of equipment reference should be made to the manufacturers' manuals.

The energy required to generate elastic waves in the ground is produced by an impact, explosives or a mechanical source. The arrivals of the various seismic waves from the impacts are detected by geophones placed in a straight line through the impact points. The geophones transform the mechanical energy into a small electric current, which is transmitted by cables to a recording unit where it is amplified before the signals are recorded directly or the information is stored on magnetic tape.

The seismic recorder is fundamentally a device that gives the time of the impact moment and the times of the arrivals of the waves at each geophone. Figure 5.1 shows a typical record, called a seismogram, obtained by the 24-channel ABEM Trio seismic refraction system. The record is from the field example of Fig. 6.19 in Chapter 6. This particular equipment is of the light-beam recorder type with moving photographic recording paper. The electric current, i.e. the signal from the individual geophone, passes after

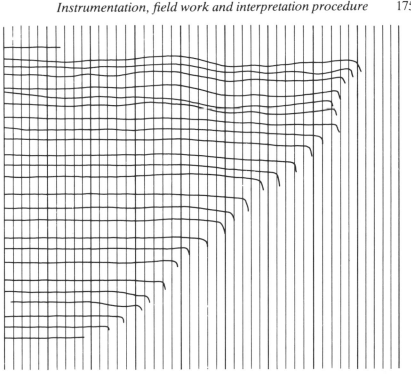

Fig. 5.1   Typical record in seismic refraction work (courtesy A/S Geoteam, Norway).

amplification through a galvanometer which is equipped with a small mirror. Before the geophone is set vibrating due to the seismic waves, a light beam reflected from this mirror records a more or less straight trace on the moving photographic paper. When the current from the geophone passes through the galvanometer, the mirror is deflected to the side and the trace on the paper no longer remains straight. On the seismogram of Fig. 5.1 the uppermost horizontal line has a break that gives the shot instant. Traces 2 to 25 belong to the 24 geophones. The vertical timing lines across the record are produced in this equipment by a light beam through a slotted rotating drum. The speed of rotation is synchronized with the speed of the recording paper so that the time interval between two adjacent lines corresponds to 2 ms. Every fifth line corresponding to a time interval of 10 ms is thicker.

In recent years instruments have become available for refraction work in which the signals from the geophones and other information are stored on magnetic tape. The data collected on the tape can be viewed on a cathode ray screen display for monitoring and time reading but the times and the timing lines can also be printed out in the form of an ordinary seismogram. Instruments equipped with digital memory enable stacking of more than one signal from the impact point, i.e. the signals are added to each other so that a series of identical weak signals can produce a strong final signal. At the same time, random background noise caused by other ground vibrations is minimized on the final record. These modern instruments generally have frequency filtering options so that certain frequencies can be discriminated to obtain selective information.

The amplification of the signals can be arranged for a particular geophone channel or for a group of channels. The gain can be controlled manually or monitored by automatic gain control, AGC, combined with automatic noise pre-suppression which reduces the gain until the first signal from the impact is received. The amplifier gain has to be kept low for the geophones near the impact point and increased with increasing distance from the impact. The general rule is to try to keep the amplification as high as the background noise allows in order to get sharp signals.

The detectors used on land, referred to as geophones or seismometers, are of the moving-coil electromagnetic type. They consist of a permanent magnet in the field of which there is a coil supported by a light spring. When the geophone is subjected to an impact, the magnet moves but the coil, because of its inertia, remains substantially in the original position. The relative movements between magnet and coil generates a voltage between the ends of the coil. The geophone ordinarily used in refraction seismics functions for vertical movements and is very little sensitive to horizontal movements in the ground. The geophones should therefore be planted into the ground in an almost vertical position. The insensitivity of the geophones to horizontal movements is sometimes observed when they are placed on outcropping solid rock. The signals from the geophones are weak because of the near-horizontal wave propagation direction.

For special measurements, for instance, shear wave and vibration determinations, three-component geophones are used. The geophones are orientated in three different directions. One geophone records vertical movements, while the other two record horizontal movements in two mutually perpendicular directions.

In water-covered areas pressure detectors, hydrophones, are frequently utilized. The hydrophones, unlike geophones, are sensitive not to movements but to the increase in water pressure caused by an impact. Hence, they are less disturbed by turbulent flows in rapids. In swampy terrain geophones of marsh type, hydrophones, or ordinary geophones mounted on top of steel rods driven down to solid ground can be used.

For multi-channel equipments the detector cables are generally joined in groups of twelve protected by a plastic cover. The distance between the geophone takeouts on cables to be used on land is somewhat larger than the nominal distance to allow for rough terrain, while hydrophone cables are manufactured with exact takeout distances. Cables intended to be used in salt water must have watertight connections at the takeouts.

5.2 FIELD WORK

In shallow refraction surveys for engineering purposes and water prospecting, the depths and velocities obtained must be accurate and give a detailed picture of the subsurface conditions. It is advisable to use the continuous profiling measuring technique with reverse recording in order to get a sufficient amount of data for a reliable analysis. The distances between impact points and between geophones have to be kept rather short.

The in-line profiling system is generally employed in refraction seismics, i.e. the impact points and the geophones are in the same line. Figure 5.2 shows a standard field set-up for 24-channel equipment. The instrument, comprising

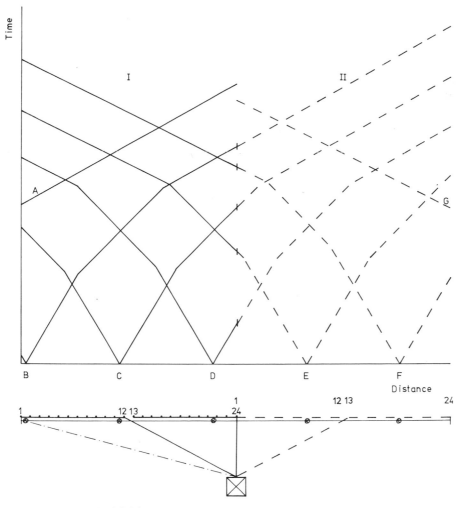

Fig. 5.2 Standard field set-up.

recorder and amplifying unit, is placed at the end of the first geophone layout/spread (I). Two geophone cables are used, one for geophones 1–12 and the other for geophones 13–24. The recording equipment is connected to the geophone cables by separate cables; the connection points are at geophones 12 and 24. The cable between the impact point B and the recorder is shown by a dash-dotted line. It is an advantage to have the seismic profile marked by, for instance, small pegs at the geophone stations in order to facilitate and speed up the field work. If the terrain is flat, a measuring tape or a line with the stations marked on it can be used. It is convenient to place the geophone cables on the other side of the seismic profile than the instrument site, as indicated in the figure. The work at the impact points can thereby be carried out without the personnel getting mixed up with the geophones and the geophone cables.

For the first geophone spread (I) in the figure, records are made at points A, B, C, D, E and F. The records from A and F provide a coverage of arrival times from the bottom layer (the third) in both directions, as can be seen in the time–distance graph. The record from E is needed for a depth determination to be made at this point. For the next spread (II), the geophones and cables are moved to the position shown by the dashed lines. The longer connection cable now runs between the recorder and geophone 13, while the short cable is connected to geophone 1. The overlapping of arrival times is obtained by means of geophone 24 in the first spread and geophone 1 in the second. Time overlapping to enable the connection of velocity segments from the various layouts can be made by one or two geophones. In the second geophone layout, records are repeated for the points B, C, D, E and F. The records from the impacts at points B and G give the travel times for the waves refracted from the third layer. The recorded arrival times of the two geophone spreads are connected to each other by means of the times at the overlapping geophones. This results in continuous curves from the impact points, and the total depths (to the third layer) can be calculated at each impact point. If in the first layout a record from E is not made, we could not calculate the depth of the third layer at this point, since in the second layout only velocity segments from the first and second layers are represented in the travel time curves and the break point between the second and third layers lies within the first geophone spread. The same reasoning is valid for the records from impact point C.

We have to keep the geophone distances rather short to obtain a sufficient number of arrival times from the various velocity layers but at the same time we may not be able to record all layers in one spread. The solution in such a case is to repeat the impacts over more than one geophone spread, an indispensable technique if we want to have detailed and reliable results.

When commencing the recording work at a site where the geology and conditions for the energy transmission are unknown, it is advisable to make a test record before the measuring procedure is decided on. Assume that this is to be made at point B in Fig. 5.2. The energy source may be mechanical in nature, or explosives. The test is performed to discover the amount of explosives or the number of mechanical impacts needed to obtain signals yielding an acceptable accuracy in the time reading. The record also gives indications of the number of velocity layers composing the subsurface, three layers in this case. The information obtained from the trial record can be used to fix the measuring procedure for the first layout, estimate the energy needed, the distance beyond the spread for the offset impact A to have registrations from the bottom layer, extent of repeat recording, i.e. the impact points in the next layout, in this case E and F.

When the records A–F have been completed, further information can be extracted from the travel time curves displayed on the seismograms. An estimate of the velocity in the third layer is obtained by comparing curves A and F. The parallelism between the various curves is used to estimate the distances between the impact points and the break points. A rough depth calculation can be made, since the depth in general lies between a third and a fifth of the critical distance, i.e. the distance from impact point to break point.

The above sketched analysis of the geological conditions can be made directly from the records of the first layout and the field procedures for the next layout can be settled. However, checks must be made continuously during the measurements to ensure that the offset impact points and the repeat recording are sufficiently extended in order to, as often popularly expressed, 'keep contact with the bottom refractor'. If a profile is longer than the two spreads in Fig. 5.2, it is convenient to move the recording equipment along the profile to positions from which two spreads can be measured in order to save time. The measuring technique described above may seem laborious and tedious, but with careful planning of the field work and efficient use of the seismic team a satisfactory measuring rate can be kept up.

A geophone separation of 5 or 10 m is recommended in general for shallow refraction investigations. The field work and interpretation are facilitated if a uniform distance between the geophone points is employed. Geophone distances of 5 m are used when the refractor depths do not exceed 25–30 m. For greater depths the distance is increased to 10 m while geophone separations of 15–20 m are adequate when the depths approach 70–100 m. The distance between the impact points generally varies between 25 and 50 m, when the corresponding geophone separation is 5 or 10 m. In order to avoid excessive recording work, the impact points should be 150–250 m apart when the depths approach 100 m.

The field procedure outlined above is not to be regarded as a rigid system to be used under all conditions. It has to be modified to suit the aim of the actual project and the geology. In the case of a survey that is more in the nature of a reconnaissance, the distances between geophone stations and/or impact points may be increased while they may be decreased if very detailed information is needed for the project. In the latter case it is not uncommon to

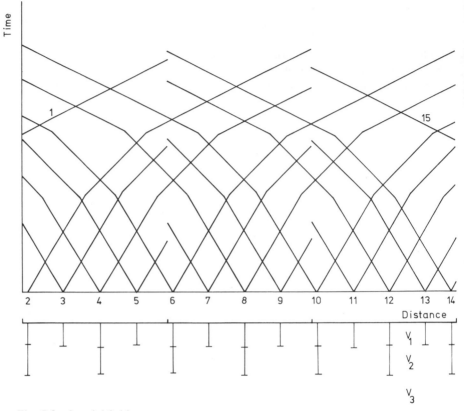

Fig. 5.3   Special field set-up.

reduce the geophone separation to 2.5 m and simultaneously make use of two sets of equipment to maintain an acceptable measuring rate. There are also combinations. A reconnaissance survey can be followed by detailed measurements with reduced impact and geophone distances at sections that are critical or particularly interesting for the project.

The design of a project and the actual geology are sometimes confounded when planning measurements. If, for instance, a tunnel is planned to be located at a depth of 50 m below ground in an area where the depths to the compact rock are 5–10 m, the performance of the field work has to be adapted to the geological conditions and not to the depth of the proposed tunnel. Therefore, short distances between impact points and between geophone stations must be employed in this case.

Sometimes detailed information on the upper layers as well as estimates of the depths and velocities of considerably deeper layers are requested. In such a case the field procedure of Fig. 5.3 may be utilized in order to reduce the time for the recording work. Inside the geophone spreads records are produced at each impact point, while outside the spreads only every second point is recorded, namely those with even numbers. The depth penetration to the third layer is achieved at the impact points with even numbers, while the depth of the second layer is obtained at all impacts, as shown in the cross-section in the figure.

The quality of the registrations, the signals, from the impacts varies greatly depending on the geological conditions. Thick layers of dry sand, silt and gravel absorb the energy effectively and it can be difficult to obtain acceptable data. On the other hand, water-saturated layers, provided they are not organic in nature, seldom create problems for the production of sharp arrivals. When explosives are used as the energy source, the charges vary considerably, from a blasting cap to some kilograms. The shots are generally fired in open pits or in holes made by means of a steel bar. In built-up areas or where it is difficult to obtain good signals, the shots can be placed at the bottom of iron tubes driven into the ground or in relatively deep-drilled shotholes. Special attention should be paid to the coupling between the geophones and the ground. Geophones equipped with spikes have to be firmly planted into the ground to increase the reception capacity. If necessary, the uppermost loose soil layer should be removed. Besides, the disturbance due to wind is reduced if the geophones are placed in small holes.

In principle, the measurements in water-covered areas are carried out in the same manner as those on land. The technique to be used depends on the size of the investigation, water depths, the accessibility of the site and the accuracy and information demanded for the project.

For detailed investigations the hydrophones and shots are lowered to the bottom. The hydrophone cable supported by buoys floats on the water surface. The hydrophones are connected to the takeouts on the hydrophone cable by separate cables. The shots are placed between the hydrophones in the same pattern as for land measurements (see Fig. 5.2). The floating hydrophone cable can be supported by an anchored floating rope or thin steel wire to reduce the stress on the cable. If the profile to be measured is longer than one spread, the rope/wire gives the position and direction of the profile and when one layout is completed the hydrophones are lifted up from the bottom, the cable is moved to the next position in the profile and the hydrophones are lowered to the bottom again. Overlapping of arrival times is arranged at one or two hydrophone positions. Repeat recording can be carried out in the same way as on land according to Fig. 5.2. The auxiliary equipment needed for this type of measurement includes some rowing-boats (perhaps equipped with outboard motors), buoys, rope/wire and anchors. Some precautions must be taken when working on water to avoid accidents. Communication between recording and shooting crews must be perfect. Before a shot is fired, the shotboat must be removed from the shotpoint at least the same distance as the actual water depth. When lowering a shot to the bottom it is necessary to make sure that the shotcable is free and has not become stuck in the boat. Some ridiculous and quite unnecessary incidents have taken place in the field because of shots dragging in the shotcable immediately behind the shotboats, and after firing, the shooting crews have found themselves sitting in the water with the oars in their hands but with no boat below them.

A more rapid way to carry out detailed surveys is to use a hydrophone cable with the hydrophones fixed to the cable, the so-called streamer type arrangement. The streamer is generally supported by a rope or a steel wire. The shotcables are either attached to the streamer or else the shots are lowered separately to the bottom. In both cases the shots have to float about 1.5 m above the water bottom in order not to damage the cable or the hydrophones. The streamer is kept in place on the bottom by attached, heavy weights. The position of the streamer is determined by means of self-adjusting buoys at each end of the streamer.

Sometimes, it is preferable to have the streamer floating close to the water surface if the water is not accessible, for instance, across rapids. The shots are placed 1–1.5 m below the water surface in order not to damage the hydrophone cable. In the interpretation the water layer is treated as an ordinary velocity layer.

Another way to measure across rapids is to reverse the shot and receiving positions. Geophones are placed on land in the same way as we use offset impact points to obtain travel time data from the bottom refractor. For the determination of overburden velocities and of depths, geophones or hydrophones are used close to the river banks. Shots are fired at regular distances across the river, replacing the ordinary receivers. The shots are placed in their position by pulling them across the river. The shots equipped with heavy weights are dropped into the river and fired before the stream carries them away.

For large-scale investigations where the refraction method can be too slow and expensive, the main part of the survey may be carried out by acoustic profiling, Boomer and Sparker, while the seismic refraction measurements are merely resorted to for solving particular problems like determining the velocities in the various layers or facilitating the interpretation of the Boomer and Sparker data.

When working in water-covered areas the positioning of the seismic profiles is generally made by intersections from two theodolite stations. The water depths can be determined by lead-line or by echo-sounding.

5.3   FIELD SHEETS

A lot of trouble can be avoided if the notes from the field work are comprehensive. The profile and spread numbers and the position of the impact points must be evident from the field sheets and from the records. It is advisable to draw a simple sketch of the position of the profiles giving the numbers, zero and end-points of the profiles and some outstanding features of the terrain. This sketch can then be used for comparison with the mapping of the profiles made by the surveyor. Discrepancies between drilling and seismic results have sometimes been due to confusions concerning profile numbers and/or zero points.

The interpretation work is facilitated if geological observations made during the field work are also noted in the field sheets. Some factors of interest are: types of soil, moisture conditions, outcrops and slope of bedrock, quality and type of rock, faults and foliation (dip and strike), areas with filling or concrete, position of drillholes and testpits in relation to the profiles. Depths of shothole and length of geophone rods must also be noted.

5.4 TIME–DISTANCE GRAPHS

The field examples in Chapter 6 are all processed in the conventional manner, i.e. by manual work on conventional time–distance graphs. The processing can be computerized, but the interpreter nevertheless needs a time plot displayed on a viewing screen and/or on a printout for an evaluation of the interpretation work.

Some points should be observed when the interpretation is made on time–distance plots. The scale chosen must allow an adequate time resolution, commensurate with the time reading accuracy of the equipment used. The relation between the time and distance scales should be such that the velocities can be obtained easily without lengthy and tiresome calculations. The scale relation for the field examples in Chapter 6 is shown in Fig. 5.4. It is

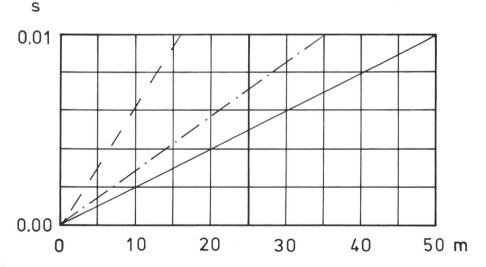

Fig. 5.4 Easy way of calculating velocities.

based on a length scale of 1:500, i.e. 1 cm on the plot represents 5 m in the terrain. A stretch of 50 m is represented in the figure. On the vertical scale 5 cm corresponds to 0.01 s. Therefore, the slope of the solid line from the zero point to the upper right corner of the figure represents a time of 0.01 s over a distance of 50 m. If the two quantities are multiplied by 100, the result will be 1.0 s over a distance of 5000 m, i.e. the velocity can be directly 'read' as 5000 m/s, simply by reading the distance of 50 m. The dash-dot line gives the velocity 3500 m/s and the velocity indicated by the dashed line is 1600 m/s, 16 m on the plot corresponding to 0.01 s. The millimetre and half millimetre marks are not shown in the figure for practical reasons. The work is further facilitated if on the millimetre paper every fifth centimetre line is a thick line. If equal multiples of both scales are employed, the relationship between the scales is maintained and the velocities can be read directly even in such a case. The length scales 1:1000 and 1:2000 can be employed when the depth of the refractor is great. On the time scale 5 cm then correspond to 0.02 and 0.04 s respectively.

When working with the time–distance plot it is sometimes necessary to change over from one curve to another or from one set of reversed curves to another set for corrections and velocity determinations. Some procedures to be used in such cases are given below.

In Fig. 5.5 the ABEM method has been used as a means of correction to obtain true intercept times. The travel time curves are only given for the forward recording. There are no problems in correcting the refractor velocity segment from point B since the breakpoint between the overburden and the refractor velocity segments occurs before the end of curve A. The time differences between curves A and B are equal at the last three geophone positions on curve A, i.e. the curves are parallel over this interval. On the other hand, in correcting the refractor arrival times from the impact point C, we have to change over from curve A to curve B at the end of curve A. Curve B cannot be used directly since the arrival times on this curve vertically above C refer to the overburden. A distance equal to the time difference between the last arrival time on curve A and the correction line (a) has to be marked off below the corresponding arrival time on the curve from B and through the point thus located a new correction line (b) has to be drawn. The correction terms for the arrival times from the point C are then obtained by taking the time differences between the times on curve B and the correction line (b). A similar problem occurs for the correction of the travel times from point E. The overlying curves from B and C end, before the waves from E are refracted back from the refractor. Observe that in this case the position of the correction line (d) has to be above the arrival time from point D with the same time difference as between the last arrival time on the curve from B and the correction line (c). It should be noted that more than one arrival time can be used when changing from one curve to another, particularly if there is no perfect parallelism between the curves owing to difficulty in plotting weak

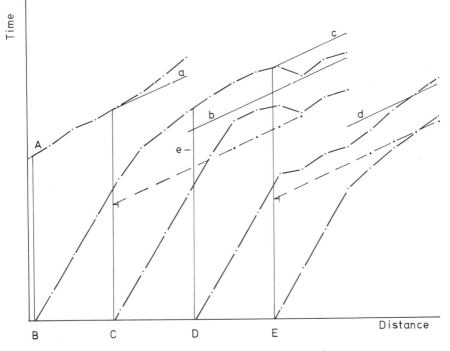

Fig. 5.5   Procedure for 'changing over' from one curve to another.

signals. One way to avoid the problem with the curve change-over procedure is to displace parallelly the refractor curves from B and D and connect them to curve A so as to form a continuous travel time curve over the entire traverse. The latter technique is, however, more time-consuming.

If the law of parallelism is used to obtain the intercept times at the impact points, there must also be a change between the curves. When correcting the refractor velocity segment from point C, the time difference between the curves from B and C is plotted at (e). The time difference between (e) and the corresponding arrival time on curve A is then projected back to the time axis through point C in the usual way.

In Fig. 5.6 the refractor velocity has been determined by the mean-minus-T method from curve 7. For layout I, the time difference $\Delta T_1/2$ has been plotted from an arbitrarily chosen reference line (a) in the usual manner. At the end of the first layout I, curves 3 and 4 have to be used for the velocity computation. The connection to the time calculation of the first layout is made by laying off a distance $\Delta T_2/2$ to obtain a point through which a new reference line marked (b) is drawn. Observe that when using curves 5 and 6, the reference line (c) must be placed below the last time point in layout II. Any mistake in placing a reference line on the wrong side will be revealed immediately when the calculated mean velocity gets a 'negative' value. The

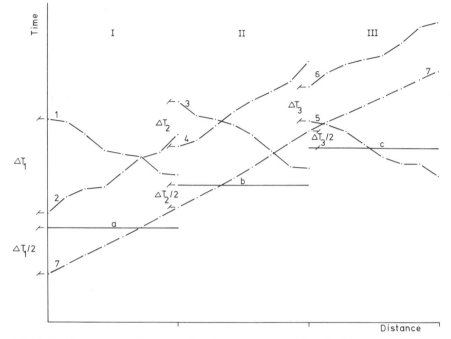

Fig. 5.6   Curve connection example in the mean-minus-*T* method.

problem of changing from one set of curves to another set can be circum-
vented by joining the velocity segments from the various layouts to one
continuous curve in each direction before the mean velocity is determined.

5.5   INTERPRETATION PROCEDURES

The execution of the interpretation is a matter of choice and no definite rules
can be laid down. Some remarks, however, may not be out of place here and
may serve as a guide. The recommendations below are adapted to the
methods described in Chapter 4.

It is vital to make the time reading very carefully. The practical accuracy
should be better than 0.5 ms for the individual arrival time. When plotting the
recorded times, special attention must be paid to the arrival times around the
breakpoint between the overburden and the bedrock velocity segments.
Inaccurate reading of the time data in this region may obscure indications of a
bottom layer, resulting in serious errors in the depth determinations, and lead
to a hidden layer problem not caused by geological conditions but produced
by the interpreter. On the other hand, in shallow refraction work the impact
moment is of less interest. Errors introduced by inaccuracies in the determination of the zero time are insignificant owing to the small distances
between the impact points and the nearest geophones.

The next step in the interpretation is to make an overall check of the time plotting. The parallelism between the various refractor velocity segments has to be checked. Departures from parallelism may indicate inaccurate time readings. The travel time curves are then examined to find the likely number of layers, the apparent velocities in them and the positions of apparent breakpoints.

An ABC-curve based on the arrival times from the bottom refractor facilitates the understanding of the subsurface conditions and can be used later on to detail the refractor surface between the impact points. Such a curve can also serve as a means to correct raw travel times.

Before commencing the depth analysis, we must know the velocity distribution in the refractor (the bedrock). It is convenient to carry out the first velocity determination by the mean-minus-T method. Referring to curve 7 in Fig. 5.6, we see that the corrected times on the curve are joined by short lines. These lines act as a guide when evaluating the velocities. An interpretation technique using the corrected times to calculate a best-fitting velocity line will lead to an average velocity and details in the velocity distribution are very likely lost. Deviations and scattering of the corrected times signify particular geological features, or sometimes a less accurate time reading. For the evaluation of the velocities it is convenient to start with the segments where the velocities are relatively high. When these are settled, the interpreter can proceed to study sections with lower velocities. The corrected times on curve 7 show higher velocity lines in the first layout extending into the second and at the end of the profile in the third layout. Between these two velocity sections, there are indications of a lower velocity at the end of the second layout. Further proof of the presence of a low-velocity zone can be obtained by prolonging the high-velocity segment in the beginning of the profile to the right. In the third layout curve 7 and the prolonged velocity line will be displaced in relation to each other, indicating a time delay due to a shear zone.

The intercept times at the impact points can be determined either by the ABEM method or by the law of parallelism. If the ABEM correction method is used, the lines for the corrected refractor (bedrock) velocity segments can be plotted immediately for each travel time curve, but it is advisable to evaluate the overburden velocities after all the corrections have been made for the entire profile. The total velocity picture, recorded and corrected travel time data, has to be considered when drawing the overburden velocity lines.

When the depths are calculated at the impact points, the refractor surface can be detailed by applying the ABC method. In the case of more than one overburden layer, a composite velocity term yielding the depths previously obtained at the impact points can be used for the depth determinations according to the ABC method.

The computed refractor velocities have to be reconsidered when the depth model is plotted. In the case of high-relief structures or when the depths are considerable, the depth calculations and/or the determination of the refractor velocities can, if possible, be renewed by Hales' method. Sections where the depths obtained seem to be doubtful can be examined more closely by means of the modified ABC method and/or the construction and displacement techniques described in Chapter 4.

5.6   PROFILE POSITIONING

Besides depending on the execution of the proper seismic work and the geological conditions, the reliability of the seismic results also depends on the positioning of the profiles. There are three factors that must be taken account of when planning a measuring programme: the type of project, the geology and the specific conditions (possibilities and limitations) for the investigation method to be used. Besides these technical aspects, there is an economic side to the problem since the funds available restrict the magnitude of a survey.

The applicability of the seismic method and the accuracy of the results increase if the programme can be modified to suit the geological conditions detected. However, this implies that during the course of the fieldwork the interpretation is carried out continuously and that preliminary reports are delivered by the geophysicist. In any case, even if there is no intention of using the results of a survey to change the programme the field data ought to be treated and analysed before the seismic team leaves the site.

In general, the quality of a seismic investigation is enhanced if there is close cooperation between the project engineer, the geologist and the geophysicist. The latter ought to use his/her professional knowledge and experience to actively propose possible modifications of the measuring programme. A preliminary planning of a particular project can assume a more definite form when the investigations gradually yield a clearer picture of the geological conditions and as a result of the replanning of the project, the measuring programme has to be revised. Below I have outlined some viewpoints concerning the planning and execution of surveys for various types of project.

An investigation for a proposed dam can be initiated by some profiles perpendicular to the dam axis. On the basis of the results from these profiles, the most suitable location for the dam can be decided and one or more profiles are measured along the proposed dam axis. If a section with inferior rock quality is detected on a profile in the dam axis, the direction and extension of such a section should be investigated more closely by parallel profiles for an evaluation of possible spots for water leakage in the future.

For underground projects such as machine-halls, oil and petrol storages, air raid shelters and tunnels the rock structure has to be detailed by profiles in different directions. The measuring procedure is a little different if the project is a cavern of limited size or a relatively long tunnel.

Surveys for the first type can be carried out in two steps as a general investigation over a relatively large area with large distances between the profiles, followed by a detailed investigation on a selected part of the area where the rock cover for the project is sufficient and the rock quality, defined by the velocities, appears to be suitable. The first phase of the survey is obviously out of the question in case the construction, for some reasons, has to be located within a small area. If the rock structure can be anticipated, the first profiles can be conveniently placed perpendicularly to the main structural features. When that system is known from the results of the first profiles, other structures can be investigated by profiles with the same direction as the main structure. In order to facilitate the structural analysis the latter profiles should be placed in such a way that they do not intersect low-velocity zones of the main structure.

In areas where thick layers of overburden cover the bedrock, the structural pattern can be revealed by the velocity distribution obtained in profiles with varying directions. If the distances between the profiles are great, it is difficult sometimes to correlate low-velocity zones on the various profiles since several solutions are possible. The problem can be solved by additional profiles placed parallel and close to the first ones where there are indications of prominent low velocities. The extra profiles can be measured in a simplified manner. If there is no need for depth determinations only two offset impacts beyond the ends of the profile will suffice to obtain the rock velocities.

The geological conditions for tunnel projects are too often studied by means of only one profile, namely in the tunnel direction. Such an investigation may, however, give a false picture of the depths and the rock quality. A dip of the bedrock surface normal to the tunnel direction cannot be revealed by a profile following the tunnel. The seismic investigation gives the perpendicular distance between the profile and the bedrock surface and without other information that distance is interpreted as a vertical depth. If the tunnel direction happens to be approximately the same as that of the rock structure, the waves will follow the compact, high-velocity rock sections while the sections with an inferior rock quality will remain undetected. The velocities obtained do not represent the real rock conditions in such a case. The profile/tunnel direction may more or less coincide with a shear zone while the velocities recorded correspond to the solid rocks surrounding the zone. The remedy is to measure short profiles perpendicular to the tunnel at regular distances or at selected places. If only the rock velocities are needed it is not necessary to carry out a complete measurement along these profiles and the records from two offset impacts will generally suffice.

Tunnel entrances located under water constitute a special problem. The seismic results, depths and rock velocities, and the mapping of the position of the profiles must be very accurate since misinterpretations may have serious consequences. An underestimate of the overburden thickness may lead to blocking of the tunnel when it is opened by the last blasting, because precautions in planning the tunnel excavation will have been made then for a lesser quantity of loose material. Moreover, the bedrock surface is encountered earlier in the tunnel than predicted if the calculated thickness of the soil cover is smaller than in reality. However, the risks of accidents are generally reduced because the geological conditions are checked by horizontal drill holes when approaching the tunnel end during the excavation work. An overestimate of the soil cover thickness may lead to a failure to open the tunnel at the last blasting, the rock cover being greater than calculated.

An investigation for a tunnel entrance starts conveniently with a profile perpendicular to the tunnel at a water depth giving approximately the level of the proposed tunnel. Profiles are then shot in the tunnel direction at a section where the rock velocity is high and the overburden is thin. An investigation for a tunnel entrance should comprise two profiles in the tunnel direction and two profiles intersecting the tunnel. In order to obtain detailed and accurate data, a hydrophone spacing of 2.5 m should be used.

The ideal site for a tunnel entrance is a compact rock massif steeply sloping at the place for the entrance and with an insignificant cover of loose material. It should be noted that a surficial layer of fractured and broken rock, indicated by lower velocities, reduces the strength of the rock cover and may cause problems in driving the tunnel.

When using seismics for water prospecting in the bedrock, the profiles are placed normal to the assumed direction of the main structural features. If the structure is unknown, profiles in various directions have to be utilized to detect possible faulted zones, fracture zones or other paths for water. When and if a promising low-velocity zone is encountered, its direction and extent are traced by additional profiles in order to locate the water well as close to the consumer as possible to reduce the installation costs.

Roads, railways and canals can be investigated by profiles covering the area of the project or by a single profile. In the latter case complementary measurements can be carried out where there are indications of bedrock or other relatively hard material that is likely to interfere with the planned excavation works.

In the field procedures outlined above an interaction between the planning of the project and the execution of the seismic survey has been suggested to obtain an optimum solution for the project at the minimum possible cost. Profiles that prove to be unnecessary can be omitted and new profiles added according to the changes in the project plan and the plan is remade in the light of the seismic results. However, there is often a tendency to have investigations carried out according to rigid, inflexible programmes (including impact and geophone distances) and the demand is to have a certain amount of seismic work completed. Such a procedure is in general against the very nature of investigations because the purpose of a survey is to investigate the unknown, and a measuring programme totally fixed in advance implies that the unknown has been already foreseen. Even when seismic measurements have been completed and results delivered according to work specifications there is always a risk that the real geological problems for the project remain undetected and the best solution for the project is not found. It must also be remembered that, from a purely seismic point of view, additional profiles in varying directions may be needed to obtain reliable results. Nevertheless, some seismic surveys can be carried out according to a programme fixed in advance, for instance, a routine investigation on a more or less grid-type profile system for the determination of thicknesses of sand and gravel deposits. Compromises between the two ways of looking at the matter are also possible as long as flexibility in the planning of the project and in the execution of the seismic work is maintained.

6

Applications

The data in the field examples presented in this chapter have been obtained by the continuous in-line measuring technique. Registrations of arrival times from the various subsurface layers have been carried out in both directions and the travel time curves from separate layouts have been connected by overlapping detectors. Since all examples given are of shallow investigations, rather small detector and shot separations have been used. The examples are selected to illustrate the application of various interpretation methods discussed in Chapter 4 to a variety of geological conditions. The first part deals simply with an analysis of the travel time curves and interpretation techniques as well as comparison between different approaches to the problems without particular attention to the true aims of the surveys. For example, an investigation that is in reality aimed at mapping deposits of sand and gravel may offer us an interesting study of rock velocities. The examples in the second part, where drilling and/or excavation data are also available, are discussed in relation to the specific project and the survey site.

Figures 6.1 and 6.2

The seismic traverse shown is a part of a longer profile. Geophone and shot separations used were 5 and 30 m respectively. For the sake of clarity, the registrations from every other shotpoint have been omitted in the figure. The offset shots A and F yield arrival times from the bottom refractor, which is in this case igneous rock.

A first analysis of the travel time curves shows that at the beginning of the traverse there are two easily detectable discrete velocity layers overlying the bottom refractor. The velocities in the uppermost layer vary between 1400 m/s and 1600 m/s and in the second overburden layer between 1900 m/s and 2200 m/s. Note that in the figure the velocities are marked in kilometres per second. At shotpoints D and E, only arrivals from the second overburden layer are evident in the travel time curves, indicating that the top layer is thinning out and ultimately disappears between points C and D.

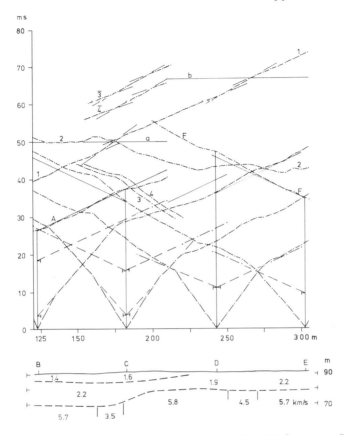

Fig. 6.1   Field example and interpretation (Archives of ABEM Company, Sweden).

Curve 1 gives the mean refractor velocities according to the mean-minus-T method. For the first part of the section, time differences between curves A and E have been measured at each geophone point, divided by two and then plotted from an arbitrary, horizontal reference line (a). The only criterion for the selection of the reference line is that the resulting mean velocity curve (1) is not tangled with any of the other curves on the graph. At the end of curve A we have to switch over to curves B and F and select a new reference line (b). The position of line (b) is determined so that a continuous refractor velocity curve is obtained. This is achieved by placing the initial point of line (b) above the last point obtained from A and E curves at a distance equal to half the distance between curves B and F at that station.

The average velocity for the relatively compact rock (curve 1) is 5700–5800 m/s. To the left of point C there is an indication of a zone with a lower velocity, namely 3500 m/s. Immediately to the right of shotpoint D the calculations also

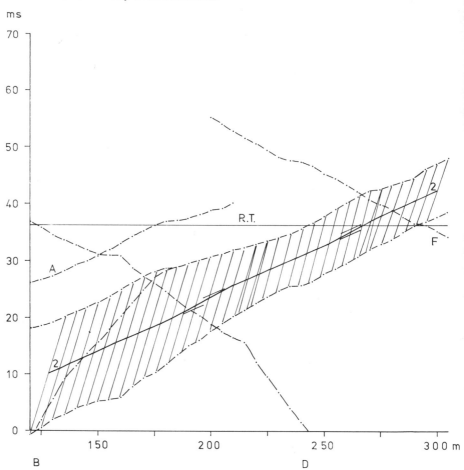

Fig. 6.2 Hales' method applied to data in Fig. 6.1.

show a velocity change in the bedrock. The velocity here is about 4500 m/s. The rock velocities are plotted on the cross-section without regard to the slant raypaths through the overburden, which means that the widths of the low-velocity zones are greater in the time plot than in reality.

The recorded arrival times from the bedrock have been corrected by the ABEM method (see Section 4.1.2). Correction lines are drawn in Fig. 6.1 above the shotpoints on the overlying recorded bedrock velocity curves. The slope of these lines has been chosen to correspond to 5500 m/s, i.e. approximately the previously calculated velocity for the solid rock. Discrepancies in the depth determination are insignificant whether we use 5500 m/s or 5700–5800 m/s. The bedrock velocity segments have been corrected but there is no need to adjust the velocity lines from the overburden layers so that the measured raw travel times have been used for the depth calculations. Curves A and B have been used for the correction of the forward curve from shotpoint C. At the end of curve A, the correction line has been applied to curve B, by making the time distances between correction lines and curves equal. For the calculation of the depths, the intercept times or the critical distances can be used according to the formulae given in Chapter 3.

Curve 2 is the ABC-curve, i.e. it represents the sum of the arrival times from the bedrock. The reciprocal time has not been considered in constructing the curve. Even if such a curve is not directly used for the depth determinations, it is always useful for an analysis of the structure. The curve also serves as an aid for checking the depth calculations at the shotpoints and may reveal errors if there is a pronounced divergence between the curve and the plotted geological picture.

In Fig. 6.2 Hales' method for the determination of refractor velocities has been applied to the travel time curves of Fig. 6.1. The time loop is closed between points B and D. The two limbs of the loop have been completed by means of curves A and F. The inclinations of the slope lines are calculated with regard to the velocities and the number of layers in the overburden. At the beginning of the profile (Fig. 6.2) the inverse slope of the lines is 750 m/s. This inclination has been calculated according to Equations (4.88) and (4.89). The first step is to calculate an average velocity V_x to replace the two overburden velocities. At shotpoint B the calculated depths are 3.4 m and 16.0 m and the corresponding layer velocities are 1400 m/s and 2200 m/s. Thus, $3.4 \times 1400 + 16.0 \times 2200 = V_x \times 19.4$, which gives $V_x = 2060$ m/s. The inclination of the slope lines is then obtained by the equation $V_x \sin i_{xn} = 2060^2/5700 = 745$ m/s. Between 225 and 275 m on the profile, where the overburden velocity is 1900 m/s, the inverse slope of 650 m/s has been used. Since we have to deal here with a simple two-layer case, the calculation of the inclination is based on the term $V_1 \sin i_{12}$. At the end of the traverse, where the overburden velocity is 2200 m/s, the value used for the inclination is 850 m/s. The centre line (2) shows the same lower velocity to the right of point D as that obtained by the mean-minus-T method in Fig. 6.1. On the other hand, the zone at point C in Fig. 6.1 has a different position in Fig. 6.2 and the velocity is higher. This result can be checked in Fig. 6.1. The bedrock velocity segment from shotpoint E is displaced 15 and 20 m to the right, as shown by curves 3 and 4. The displaced curves and the forward curves A and B are used to recalculate the mean bedrock velocity. The resulting curves $\overline{3}$ and $\overline{4}$ are then plotted half the displacement distance to the left, i.e. 7.5 m for curve $\overline{3}$ and 10 m for curve $\overline{4}$. Both curves agree with the calculated bedrock velocity curve (2) of Fig. 6.2 in indicating that there is a lower velocity immediately to the right of point C. The bedrock velocities given in the section in Fig. 6.1 refer to those obtained in curve 1 in Fig. 6.1.

Figure 6.3

The figure shows a simple two-layer case with very small velocity variations in the overburden. In the vicinity of shotpoints C and D the velocity is 1600 m/s. Elsewhere it is 1700 m/s. The straight velocity lines (overburden) indicate homogeneous conditions so that there is no need for any corrections. The recorded raw travel times can be used directly for overburden velocity determinations and for the depth calculations. On the other hand, the travel time curves emanating from the bedrock (here igneous rock) have to be corrected.

Curve 1 represents the mean bedrock velocities according to the mean-minus-T method. The velocity in the compact rock is about 5800 m/s according to curve 1. At the beginning of the profile the velocity is somewhat lower in curve 1, about 4700 m/s. Between points D and E the calculations indicate a section with a markedly lower velocity, less than 4000 m/s. The ABEM method has been employed to correct the bedrock velocity segments from the shotpoints. The slope of the correction lines applied above the shotpoints on the overlying correction curves corresponds to a velocity of 5800 m/s, except above shotpoint B where the velocity 4700 m/s has been used since curve 1 here indicated a lower velocity in the bedrock. Between the shotpoints the bedrock surface has been detailed by means of the ABC method. The depths obtained by the ABC method are marked by crosses on the cross-section. These depths are determined relative to the depths previously calculated at the shotpoints (see Section 4.1.4(b)). The time differences between the ABC-curve (2) and the horizontal lines drawn from points on the curve directly above the shotpoints yield the depth variations. Note the agreement between the computed intercept times at the shotpoints and the corresponding intercepts in the ABC-curve. It can also be observed that the ABC-curve in this particular case is the mirror image of the bedrock configuration.

It is very likely, however, that the interpretation underestimates the depths between the shotpoints D and E. There is a low-velocity zone here and the bedrock surface is dipping towards the zone. In order to study more closely the bedrock configuration in this section, the travel time curve from G is displaced 20 and 30 m to the right, giving curves 4 and 5 respectively. A new summation in this position of the direct and reverse arrival times yields the curves $\overline{4}$ and $\overline{5}$. These curves have been moved on the plot 10 and 15 m (half the displacement distances) to the left. There is a clear indication in these curves that the greatest depth is encountered between points D and E. The time differences between curves $\overline{4}$ and $\overline{5}$ and horizontal lines drawn through points on them directly above point D show that the depth between 105 and 110 m on the profile is about 16.5 m, i.e. about 2.5 m greater than that at shotpoint D. Since the curves are displaced, the proper equation for calculating the depth, without considering the dips, is $h_1 = T_2 V_1 \cos i_{12}/2$ according to Section 4.1.6(a)(ii), Equation (4.65). The term T_2 in the equation should, strictly speaking, be written ΔT_2 since we are calculating depth differences.

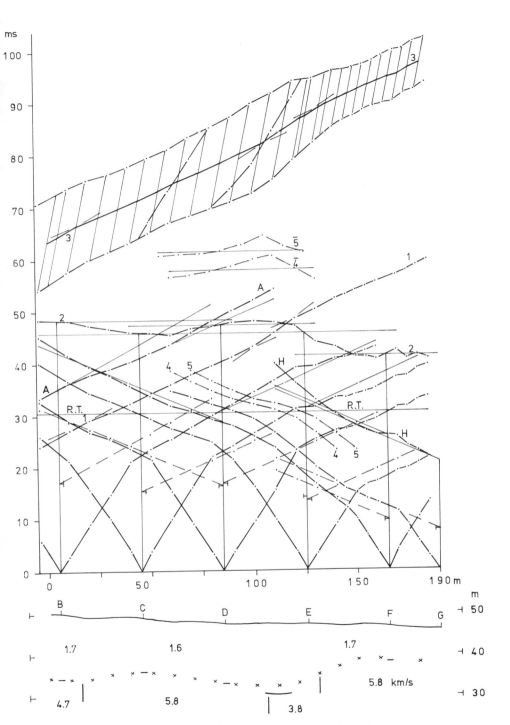

Fig. 6.3 Field example of two-layer interpretation with varying velocity in refractor (courtesy A/S Geoteam, Norway).

The maximum depth can also be estimated by using the original ABC-curve and the equation $h_1 = T_2 V_1/2\cos(\varphi_2 + i_{12})$ according to Section 4.1.6(a)(i), Equation (4.60). Note that T_2 in this case denotes the time difference between the ABC-curve and the reciprocal time. Employing this technique we obtain a depth of 17.0–17.5 m. The calculated maximum depth is marked on the section by a short arc between the shotpoints D and E.

The refractor velocity analysis according to the mean-minus-T method is confirmed by curve 3 which is obtained by Hales' method. The relatively low velocity at the beginning of the profile and the much lower velocity near point E are present in both curves. However, the horizontal positions differ. The velocity boundaries plotted on the cross-section are those indicated by curve 3.

Figures 6.4 and 6.5

Figure 6.4 shows a three-layer case with a depression in the bedrock in combination with a rock section with lower velocity in the central part of the profile. The velocities in the uppermost layer vary between 500 m/s and 700 m/s. The velocities 1500–1600 m/s in the second layer are obtained from the mean-minus-T curves 3–7. The lower velocity 1500 m/s is found in the vicinity of shotpoints C and D. The determination of the mean bedrock velocity from curve 1 shows a low velocity, approximately 3100 m/s, between shotpoints D and F. To the left of this rock section the velocity in curve 1 is 4900 m/s and to the right 5600 m/s.

For the depth determinations outside the low-velocity zone the ABEM correction method has been used in the usual manner. At shotpoints B, C and D the correction lines have the inverse slope of 5000 m/s. At the end of the profile, points F, G and H, the slope of the lines corresponds to the velocity 5500 m/s. A relative ABC-curve, namely curve 2, has been plotted as an aid for the depth analysis.

It is very likely that the depths in the vicinity of point E will be under-estimated owing to the combination of a low-velocity zone and a depression in the bedrock. A conventional use of the law of parallelism or the ABEM correction method yields the intercept times marked (a) at shotpoint E. The corresponding depth is given by the arc (a) in the cross-section. For the depth calculation the velocities 1600 m/s and 3100 m/s have been used for the overburden and bedrock respectively. Approximately the same depth will be obtained at point E if within the low-velocity zone the correction lines are based on 3100 m/s and then changed to 5000 m/s or 5500 m/s at points corresponding to the velocity changes in curve 1. The correction lines, intercept times and depth are marked (b). The velocities 1600 m/s and 5000 m/s or 5500 m/s have been used for the calculation of the depth.

The intercept time (c) refers to the case when construction lines, also marked (c), are used for the correction. The correction technique is described in Section 4.1.3 in connection with Figs 4.22 and 4.23. In the direct recording

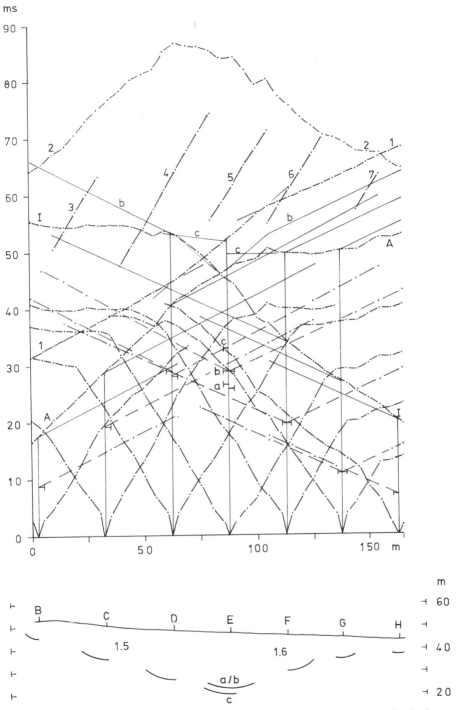

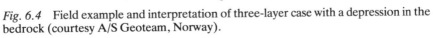

Fig. 6.4 Field example and interpretation of three-layer case with a depression in the bedrock (courtesy A/S Geoteam, Norway).

the same intercept will be obtained as for (b), while the intercept time is considerably greater in the reverse recording. The depth (c) calculated from the latter intercept value is plotted towards the left flank of the depression.

The depths calculated at shotpoints D and F are probably acceptable. The rays from the shotpoints strike the bedrock outside the low-velocity zone and curves A and I used for the corrections refer to the compact rock below points D and F. Therefore, it is likely that there is a parallelism between the curves involved in the depth determinations.

The bedrock velocity analysis in Fig. 6.5 (Hales' method) confirms the low-velocity zone of Fig. 6.4. However, the width of the zone is smaller. The inclination of the slope lines in Fig. 6.5 has an inverse of magnitude 450 m/s. It has been possible in this case to compute the velocity segments by means of Hales' method without any correction to the measured raw travel times. The depths plotted in Fig. 6.5 have been obtained using the ABC method. In order to facilitate a comparison, the depths are calculated and plotted at the locations of the shotpoints in Fig. 6.4. The two methods give almost identical results, which is to be expected since they are, in principle, based on the same concepts. It should be noted that the ABC-curve (2) in Fig. 6.5 is composed of travel times from the upper as well as the second layer. Therefore, the depth calculation ought to be based on an average velocity, i.e. a velocity somewhat lower than that in the second layer. In this case, since the uppermost layer is rather thin the error will be negligible if we use 1500 m/s and 1600 m/s for the depth calculations without any reduction in the magnitude of the velocities. The ABC method gives the same depth above the low-velocity zone as that obtained at shotpoint E when the depth calculation is made in a conventional manner. Probably a better estimate of the overburden thickness can be obtained from the formula $h_1 = T_2 V_1/2\cos(\varphi_2 + i_{12})$, the modified ABC method in Section 4.1.6(a)(i). The greater depth thus obtained is marked by the lower arc in the cross-section in Fig. 6.5. The angle i_{12} in the equation above is based on the velocity values of 1600 m/s and 3500 m/s.

Because of the slopes of the bedrock, the velocities for the compact rock sections are too high in Fig. 6.4 and too low in Fig. 6.5. The computed rock velocities in Figs 6.4 and 6.5 have to be multiplied or divided by the cosine of the dip angles. The adjusted rock velocities are given in the cross-section in Fig. 6.5.

Figure 6.6

In the previous examples it was not necessary to adjust the travel time curves from the overburden layers. It was sufficient to correct only those from the bedrock. Unfortunately, we encounter such ideal conditions rather seldom, since even small variations in elevation and thickness of the upper layers may distort the travel time curves if the material is loose and dry.

In Fig. 6.6 the overburden consists of water-saturated sand and gravel, velocities 1600–1650 m/s. The groundwater table lies 0.5–1.0 m below the

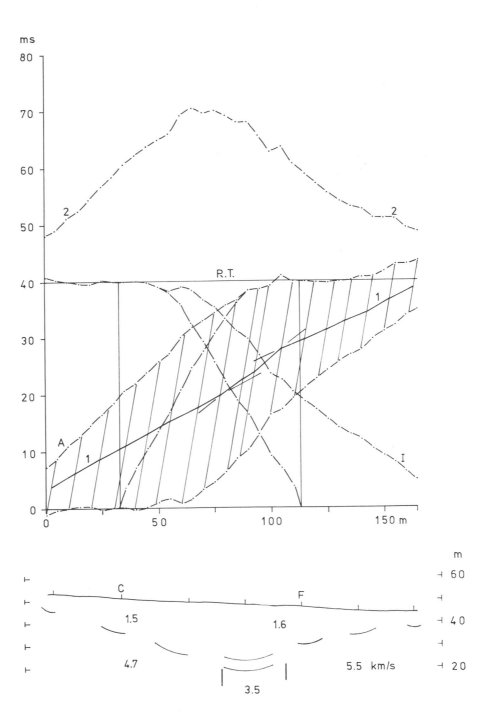

Fig. 6.5　Hales' method applied to data in Fig. 6.4.

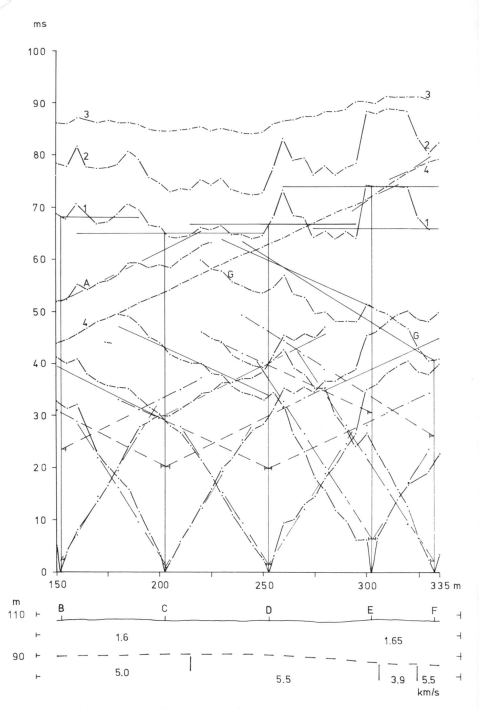

Fig. 6.6 Distorted time–distance curves (due to near-surface irregularities) and their interpretation (courtesy A/S Geoteam, Norway).

surface. Particularly at the end of the profile the lower velocities in the upper layer of dry sand and gravel have influenced the recorded arrival times considerably, despite the fact that it is very thin.

The ABC-curves for the overburden and bedrock are shown on top of the time plot in curves 1 and 2 respectively. The two curves display a similar pattern, indicating that the irregularities in them are mainly caused by conditions close to the ground surface. It may seem strange that it has been possible to construct a continuous ABC-curve for the overburden layers since there is no velocity coverage of this layer in the time–distance graph. The original data have complete coverage, however, but the registrations from every second shotpoint have been omitted in the figure for the sake of clarity.

The mean rock velocities shown in curve 4 have been calculated by the mean-minus-T method. At the beginning of the traverse the velocity is 5000 m/s. From 190 m in the profile the velocity is 5500 m/s, except for a lower velocity (3900 m/s) between shotpoints E and F. The low-velocity zone is plotted on the section without considering the slant raypaths through the overburden, i.e. with the same width as that obtained in curve 4.

The ABEM method has been used for correcting the bedrock velocity segments from the shotpoints. Correction lines with the inverse slope of 5000 m/s were applied above shotpoints B and C on the overlying bedrock velocity curves while at shotpoints D and E they are based on the velocity 5500 m/s. The correction for the curve from F and the corresponding depth calculation are a little more complicated. A critical ray from F strikes the bedrock near the boundary between the rock sections with the velocities 5500 m/s 3900 m/s and the choice of 5500 or 3900 m/s for corrections and depth calculations affects the cosine factor in the formula. The correction line used is based on 5500 m/s up to the velocity boundary in the bedrock and after that on 4000 m/s. The velocities 1650 m/s and 4000 m/s are to be used in the formula for the depth.

The overburden velocity curves have been corrected by means of curve 1, which is the ABC-curve from the same layer. Horizontal reference lines were drawn from points on this curve vertically above the shotpoints, as can be seen in the plot. The time differences between curve 1 and the reference lines are used as a correction means. Note that curve 1 gives the double up-times, so that the time differences have to be divided by two. The correction technique is described in Section 4.1.4(c).

Curve 1, the ABC-curve for the overburden layers, can be used to eliminate the effect on curve 2, the ABC-curve for the bedrock, of irregularities close to the ground surface. The relative times in curve 1, with respect to a horizontal reference line, have been subtracted with due regard to their signs from the times in curve 2. The resulting times are plotted as curve 3, whose variation is seen to be approximately the mirror image of the variation in the bedrock surface. The curve can be used for depth calculations, for instance, to map in detail the bedrock configuration between the levels obtained at the shotpoints.

Figures 6.7, 6.8 and 6.9

In this case too the recorded travel time curves are distorted because of varying conditions in the upper parts of the overburden. The same irregularities can be seen in travel time curves from the overburden as in the curves from the bedrock. The identical horizontal positions of the features on the various curves clearly indicate that the cause of the irregularities lies close to the ground surface. It can also be seen in Fig. 6.7 that there is a close resemblance between the shape of the recorded curves and that of the ground surface. At shotpoints B and D there is a top layer with a low velocity, 600 m/s, while this layer is missing in the lower parts of the terrain at points C and E. The existence of this layer at shotpoint F is uncertain. There are no records of the arrival times close to the shotpoint because of two 'dead channels'.

Different interpretation and correction techniques have been used and are compared below.

The bedrock and overburden velocities have been evaluated in Fig. 6.7 by the mean-minus-T method. For the depth determinations the bedrock as well as the overburden velocity segments have been adjusted by applying the ABEM method to the recorded arrival times from the bedrock. Curve 1 gives the mean bedrock velocities. At the beginning of the profile the computed velocity in curve 1 is about 4000 m/s. Between coordinates 10 and 90 m the velocity is 5000 m/s and at the end of the profile 4700 m/s. The calculations according to the mean-minus-T method of the velocities in the overburden, i.e., in the layer immediately below the top layer at B and D, are to be found in curves 2 and 3. This velocity is 1800–1850 m/s, except between 35 and 50 m on the profile where it is distinctly lower, namely 1650 m/s. No data are available at the end of the traverse for determination of the mean overburden velocity.

For shotpoints B–E the inverse slope of 5000 m/s has been used for the correction lines. At the last shotpoint, F, the correction is based on the velocity 4700 m/s detected in curve 1. The velocity curves from the bedrock as well as those from the overburden have been corrected by time differences between these lines and the recorded arrival times from the bedrock, namely curves A and G. As mentioned previously, the technique of adjusting a velocity curve by means of travel times from an underlying layer is not mathematically correct. In the interpretation the velocity lines for the second overburden layer are drawn using 1800 m/s as an average velocity. The positions of these latter lines are based on the corrected times, the original recorded arrival times and the mean overburden velocities obtained from curves 2 and 3. The interpretation has been simplified, by the use of the average velocity 1800 m/s, since the depths are rather small. As can be seen in the time plot, the adjusted times for the overburden (second layer) deviate from the selected velocity lines, particularly at the end of the traverse.

As a comparison, the arrival times from the bedrock and overburden have also been corrected in Fig. 6.8 using ABC-curves from the various layers. The

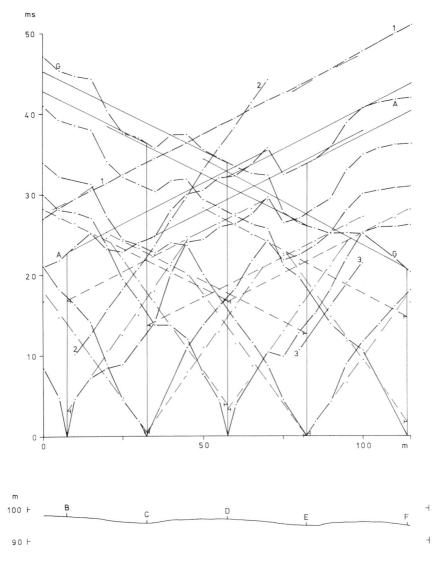

Fig. 6.7 Heavily distorted time–distance curves and their interpretation by the ABEM method (courtesy A/S Geoteam, Norway).

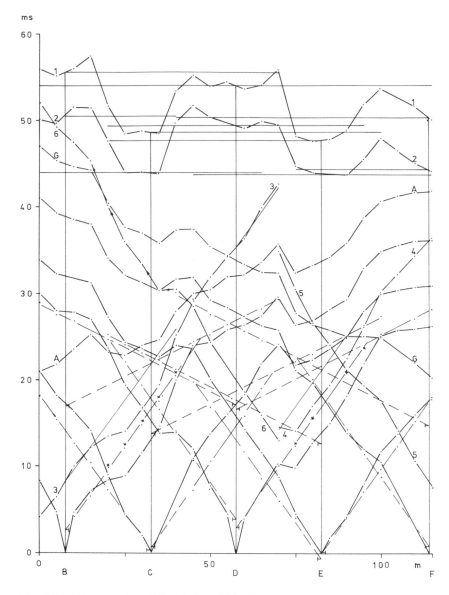

Fig. 6.8   Interpretation of data in Fig. 6.7 by the ABC method.

ABC-curves for the bedrock and overburden are shown as curves 1 and 2 respectively. As mentioned above, there is no complete velocity line coverage for the overburden. There is a gap to the left of point E, and further, no time data are available at the profile end in the reverse recording. The problem can be solved by assuming the velocity in the overburden to be 1850 m/s and constructing a reverse curve satisfying this velocity assumption by means of the direct curves. The simulated travel times, based partly on the construction and partly on the curve from shotpoint F, are shown as curve 5. The recorded forward arrival times and the simulated times in curve 5 have then been used to complete curve 2 in this region. The errors introduced by this construction of travel times are probably insignificant since the velocity in the second overburden layer is rather uniform. In the usual manner, the time correction terms have been obtained as the time differences between the ABC-curves and horizontal lines drawn from the intersections between time axes through the shotpoints and the ABC-curves. The correction technique is described in Section 4.1.4(c). The bedrock velocity lines have been adjusted by means of curve 1 and the overburden lines by means of curve 2. The corrected travel times are shown by dots. Note that the time differences between the correction lines and the recorded arrival times in curves 1 and 2 have to be divided by two before they are applied as correction terms to the curves from the shotpoints.

The two different correction techniques in Figs 6.7 and 6.8 yield almost identical intercept times for the bedrock segments. The reason is that the velocities in the bedrock are rather uniform. If there are appreciable velocity variations, the ABEM method is preferable as a means of correction. On the other hand, the corrected times from the overburden present a more regular pattern in Fig. 6.8 than in Fig. 6.7 and they define more accurately the slopes and positions of the overburden velocity lines.

The ABEM method was applied above to the bedrock velocity curves to obtain correction terms for the overburden velocities. A more accurate way is to use this method directly on the overburden velocity lines. Curves 3, 4, 5 and 6 in Fig. 6.8 are composed of overburden velocity segments (second layer) tied to each other and the correction technique is demonstrated on the forward curves points B and D. A correction line with the inverse slope of 1850 m/s is drawn from a point on curve 3 vertically above B. The correction terms, i.e. the time differences between the correction line and the recorded arrival times on curve 3, are then added with due regard to their signs to the overburden travel times from B. The corrected times are plotted as crosses. When adjusting the travel times from point D, the correction line applied to curve 3 has to be switched over to curve 4 at the end of curve 3.

The calculated depths are shown in Fig. 6.9. That the different correction techniques result in more or less the same depths without any appreciable differences is due to the fact that there are no significant velocity variations in the bedrock and that its surface is rather flat and horizontal.

Since the travel time curves are very irregular, Hales' method cannot be used in a simple way for an evaluation of the bedrock velocities. The influence of the dry low-velocity layer at the ground surface must be eliminated, or at least minimized, which can be achieved by correcting the arrival times from

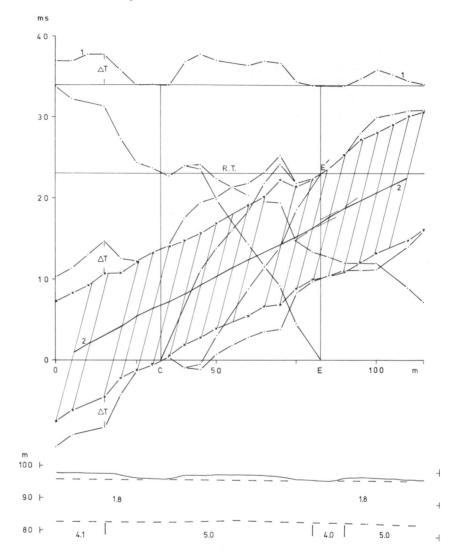

Fig. 6.9 Velocity analysis of data in Fig. 6.7 by Hales' method.

the bedrock by an ABC-curve based on arrival times from an overlying layer. The corrections cannot be perfect since the raypaths for the waves from the bedrock differ from those from the overburden.

In Fig. 6.9 the velocity distribution in the bedrock is analysed by Hales' method. The time loop is closed between C and E'. Curve 1 is the ABC-curve for the second layer of the overburden. It corresponds to curve 2 in Fig. 6.8 but the times have been divided by two. In order to maintain the reciprocal time between points C and E, the travel time from C to E must be left unchanged and vice versa for the travel time from E to C. This condition can be fulfilled if, when adjusting the direct curve, a horizontal reference line is drawn from a point on the ABC-curve (1) vertically above E and corrections are made in relation to this line. The reverse recording is corrected in the same way, i.e. the reference line is drawn from a point on the ABC-curve vertically above C. As can be seen on the graph these two reference lines happen to coincide since both shotpoints lie below the thin top layer. The application of the correction terms, i.e., the time differences between the ABC-curve and the reference line, is exemplified by ΔT to the left of point C. Since the times on the ABC-curve lie above the reference line, the correction terms have to be subtracted from the times on the limbs of the time loop. When the times on an ABC-curve lie on both sides of the reference line, the correction terms have to be added or subtracted according to whether they are below or above the reference line respectively.

The velocity values 1800 m/s and 5000 m/s have been used in the term $V_1 \sin i_{12}$ in calculating the inclination (650 m/s) of the slope lines between the two limbs of the time loop. The general velocity picture is the same in Figs 6.9 and 6.7, but the velocity 4700 m/s in Fig. 6.7 appears in Fig. 6.9 (curve 2) as a lower velocity 4000 m/s at point E and as a higher velocity 5000 m/s at the end of the profile. In the central part of the profile where the rock velocity is evaluated as about 5000 m/s there are indications of three velocity segments in Fig. 6.9 corresponding to 5200 m/s, 4700 m/s and 5200 m/s. The same tendency can be observed in curve 1 in Fig. 6.7 but it is less pronounced here. However, in the cross-section the average velocity 5000 m/s is plotted, since to divide into three velocity sections is probably to overestimate the accuracy of the method.

The cross-section in Fig. 6.9 is based on the depths obtained in Figs 6.7 and 6.8 and the rock velocity analysis in Fig. 6.9.

Figure 6.10

A first analysis of the time–distance relations shows that we have to deal with a three-layer case. At the beginning of the profile the three separate velocities are easily detected but to the right of point D the travel time curves are heavily deformed so that the velocities and the position of the velocity lines on the plot are doubtful. Since there is no coverage of registrations from the second layer, corrections of the overburden velocity curves also have to be made, using the travel time data from the bottom refractor, namely the bedrock.

The velocity analysis of the bedrock is shown in curve 1, obtained by the mean-minus-T method. The velocity in the top layer varies between 400 m/s and 600 m/s. The travel time curves from the second layer overlap short sections of the profile so that an estimate of the velocity can be made. The mean-minus-T method has given lines 2, 3 and 4 when applied to registrations from B and C, C and D, and D and E respectively. The velocity according to

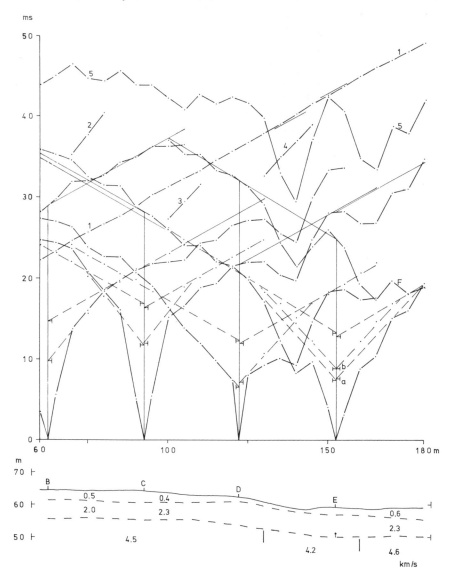

Fig. 6.10 Field example and interpretation of heavily distorted time–distance curves in a three-layer case (Archives of the ABEM Company, Sweden).

line 2 is 2000 m/s but increases to 2300 m/s to the right of point C. These rather high velocities in the second layer correspond to a compact, water-saturated moraine. The summation of the arrival times from the bedrock (the ABC-curve) is shown in curve 5.

The corrections for the depth determinations have been obtained by the

ABEM method. Since the overburden travel time data are corrected by means of arrival times from the underlying layer we can expect uncertainties. The inverse slope of 4500 m/s has been used for the correction lines. The velocity 2000 m/s has been used for the velocity lines for the second layer at shotpoints B and C and 2300 m/s at shotpoints D and E. At point C an alternative is to use 2300 m/s for the forward recording since curve 3 has indicated this velocity. At point E the upper velocity lines for the second layer, denoted by (b), are drawn through the corrected times shown by the dots. This interpretation may underestimate the thickness of the second overburden layer and, consequently, overestimate the thickness of the top layer. After such a first interpretation, the influence of the reduced thickness of the top layer on both sides of the shotpoint E can be evaluated. The position of the velocity lines (a) is based on such an evaluation. The depth-to-bedrock shown on the cross-section refers to the alternative interpretation (a) and the arrows indicate the depth obtained when intercept time (b) is used for the calculation.

Figures 6.11 and 6.12

The profile in Fig. 6.11 is measured across a valley. The bedrock is out-cropping immediately beyond both ends of the profile. The relief of the terrain has been formed partly by the small brook at point F by erosion of the soil layer to the right of shotpoint D.

The various velocity lines on the time–distance plot in Fig. 6.11 are heavily distorted. The arrival times in curves A and I of waves emanating from the bedrock show that the overburden depth increases continuously from the beginning of the profile and reaches a maximum around shotpoint D. To the right of this point the depth decreases and is less than 1.0 m at the brook and further towards the profile end. In order to facilitate the interpretation, another profile was measured more or less perpendicularly to the profile in Fig. 6.11. The data from the additional profile are to be found in Fig. 6.12. The intersection point between the profiles is marked by arrows on the cross-sections. In fact, these two profiles are not the only ones measured at this site. The measuring programme actually comprised a series of profiles across and along the valley.

The ABEM method applied to the bedrock travel time curves has to be used to adjust the various recorded travel times since there is no coverage of velocities from the overburden layers. The choice of slopes for the correction lines is based on the mean rock velocities given by curves labelled (1) in both figures. The most convenient way to solve the interpretation problem is to commence with the profile in Fig. 6.12 along the valley. In Fig. 6.11 it is possible to discern two layers overlying the bedrock, but the magnitude of the velocity in the second layer can be determined more accurately from the data in Fig. 6.12. The corrected times from shotpoints D–F in Fig. 6.12 indicate a velocity of 2000 m/s for the second layer. This layer is partially hidden,

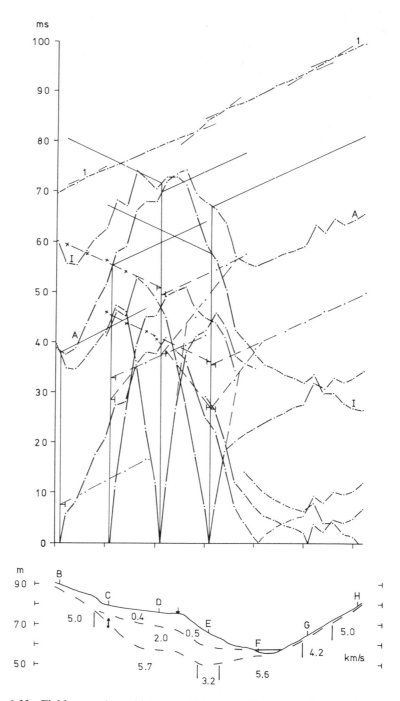

Fig. 6.11 Field example and interpretation of heavily distorted curves due to conditions in a valley (Archives of the ABEM Company, Sweden).

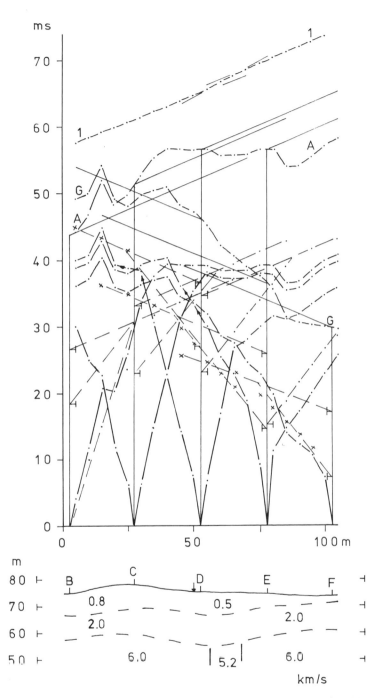

Fig. 6.12 Profile measured perpendicularly to profile in Fig. 6.11 (Archives of the ABEM Company, Sweden).

however. At the beginning of the profile the travel time curves show only the lower velocity in the top layer and the velocity in the bedrock. A calculation at shotpoint B based on the velocities 800 m/s and 6000 m/s will grossly underestimate the depth of the bedrock. Adjacent profiles indicate that the second layer, 2000 m/s, exists over the entire area. In order to obtain an idea of the maximum depth at point B, a velocity line (2000 m/s) has been placed at the intersection of the lines for the top layer and the bedrock. The depth calculation is carried out using the ordinary formula for a three-layer case. The interpretation procedure has been described in Section 3.7.

The velocity values and depth at the intersection between the two profiles obtained in the profile of Fig. 6.12 have been used in combination with the corrected arrival times to calculate the depths along the profile across the valley. The corrections show that the depth to the bedrock is only a few metres at the beginning of the profile. If these corrections are not applied the conclusion could be drawn that the depth is considerably greater. The apparent bedrock velocities in the direct recording are very low here and lie within the range of soil velocities. At shotpoint C we once again encounter the problem of the partially hidden layer. The second overburden layer is thinning out towards the left, but to what extent? The layer is not evident in the travel time curve from C. As in the case in Fig. 6.12, a line with the inverse slope of 2000 m/s has been drawn at the intersection of the corrected velocity lines of the top layer and the bedrock. The arrows on the bedrock surface below point C indicate the estimated uncertainty in the depth evaluation.

Some objections could perhaps be raised about the planning of the survey. For example, why measure across the valley when the relief is so strong and obtain data that are difficult to interpret? Of course, a large number of profiles along the valley where the terrain variation is small will facilitate the geophysicist's work, but, very likely, information that is vital for the project may remain undetected, for instance, information such as the occurrence of depressions or shear zones in the bedrock. Besides, the various correction techniques proposed enable the interpreter to solve very intricate interpretation problems caused by greatly varying topographical and geological conditions.

Figures 6.13, 6.14 and 6.15

The profile is partly in water-covered area and partly on land. Hydrophones were used for the measurements in the river and geophones for those on land. Immediately to the left of the profile the bedrock outcrops and rises very steeply. The sudden jumps in arrival times at the transition from river to dry land are probably caused by very loose intermediate layers of organic material. The soil layers on land were observed to be very loose and showed evidence of having been swampy. In the river the overburden consists of sand, gravel and boulders.

The law of parallelism has been used in Fig. 6.13 to establish the intercept

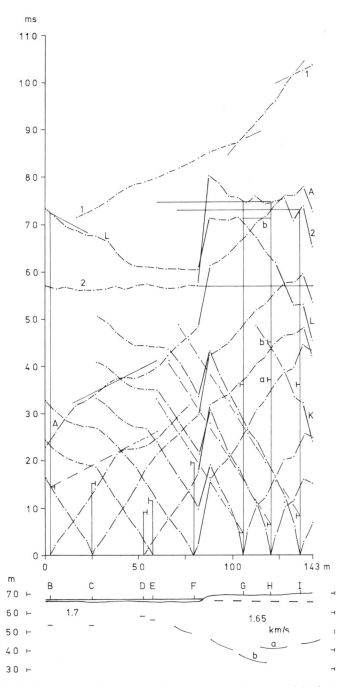

Fig. 6.13 Refraction profile in a partly water-covered area and its interpretation (courtesy A/S Geoteam, Norway).

times of the bedrock velocity lines, except for the curve from shotpoint B. The arrival times from the offset shot A shown at the very beginning of the profile emanate from the overburden layer and not from the bedrock. Note the parallelism between curves A and B in this region. From the fourth hydrophone onwards the waves from the bedrock overtake and the parallelism between curves A and B ceases. The steep slope of the bedrock to the left of the profile (not shown) probably continues below the river. The first 15 m of the profile therefore constitute a shadow zone, i.e. it is not possible in the forward recording to get any registrations from the bedrock. In such a case, the arrival times obtained from the opposite direction can be utilized, in combination with a modification of the ABEM method, to get at least an estimate of the depth. Above shotpoint B, a correction line has been applied to curve L. The slope of the line corresponds to the average velocity in the compact rock of the area. The correction line is transferred to curve A at the fourth hydrophone where the bedrock velocity segment of curve A begins. The arrival times in curve B are then corrected in the usual manner. Errors may be introduced by this technique of correcting arrival times with the help of registrations from the other direction since an assumption is involved that the refractor velocity is constant.

The overburden velocities from shotpoints G, H and I in the reverse shooting have been corrected by the ABC-curve (2), which is a relative summation of the arrival times from the overburden (second layer on land). There is a discrepancy between the time jump in curve 2 at the river bank and the intercept times for the top layer at the shotpoints. This is probably due to the fact that the shots are buried in the ground while the geophones are planted on the ground surface.

The mean refractor velocity determination according to the mean-minus-T method, curve 1 in Fig. 6.13, indicates an exceptionally low velocity between points G and I in association with the depression in the bedrock surface. It seems likely that this calculated low velocity is due to the combined effect of a shear zone and dip changes in the bedrock. It is also very likely that the depth marked (a) at shotpoint H is underestimated when the law of parallelism is applied to the recorded curves. The waves from point H strike the up-dip part of the bedrock to the left of the low-velocity zone, while curve L, above point H, corresponds to the right flank of the depression and to the shear zone in the bedrock. The parallelism between curves H and L is lost, probably immediately to the left of shotpoint G. In order to obtain an alternative depth estimate based on these assumptions, the up-dip part of curve L is projected back to the time axis through shotpoint H, line (b) on the time plot. If the law of parallelism is applied to this construction a new intercept time (b) is obtained. The corresponding depth (b) has to be plotted perpendicularly to the left side of the depression.

An interpretation using the ABC method (Fig. 6.14) gives an almost

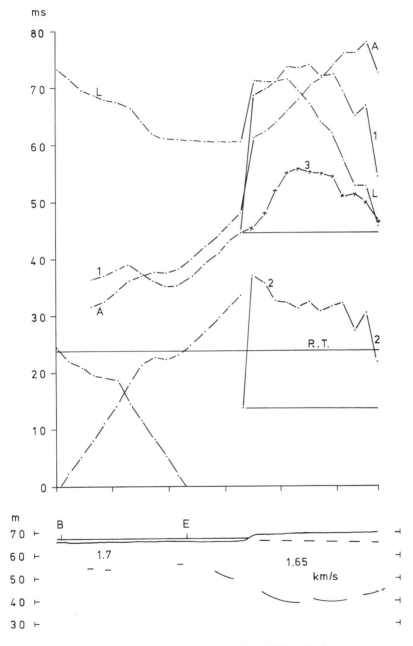

Fig. 6.14 Interpretation of data in Fig. 6.13 by the ABC method.

identical configuration of the bedrock as in Fig. 6.13. The depths in Fig. 6.14 are calculated at the shotpoints to facilitate a comparison, although in reality we use more points on an ABC-curve to detail the shape of the refractor surface. The ABC-curve (1) for the arrival times from the bedrock is based on the travel time curves from points B and E in Fig. 6.14, supplemented by curves A and L. Just as in the ABC-curve for the overburden layer in Fig. 6.13, there is a considerable time difference at the transition between water and dry land. A simplified interpretation technique has been used to calculate the thickness of the second overburden layer on the land part of the profile. The 'up-times' in curve 2, the ABC-curve for the overburden, are subtracted from the corresponding values in curve 1. The resulting curve (3) is marked on the graph by crosses. The time differences between the latter curve and the reciprocal time R.T. are used to compute the thickness of the second over-burden layer. Note that, in this case, the depths have to be plotted from the bottom of the top layer. Errors introduced by this simplified interpretation procedure are insignificant compared with the general uncertainty in the bedrock configuration in the deeper parts of the depression.

The next step for a better understanding of the geological conditions in the depression is to investigate the velocity distribution in the bedrock more closely by Hales' method. However, that technique cannot be applied to the problem without removing the influence of the travel times through the upper layer on the land section of the profile. The velocity analysis and depth calculations according to Hales' method are given in Fig. 6.15.

Curve 1 in Fig. 6.15 shows the sum of arrival times in both directions of waves emanating from the second overburden layer and corresponds to the ABC-curve (2) in Fig. 6.13. The interpretation technique has been simplified in this case also. The time increments in curve 1, because of the passage through the top layer, have been subtracted from the two limbs of the time loop. The corrected times are marked by crosses. Note that the time differences between curve 1 and the reference line R.L. have to be divided by two. The centre line (2) through the midpoints of the slope lines indicates that there are two zones with lower velocities in the depression. These velocities are about 3100 m/s and 2400 m/s and the sections are separated by a rock section with a velocity of 5500 m/s. The mean-minus-T method gave in Fig. 6.13 a single wide zone (velocity about 2100 m/s).

As expected, the depths in the depression calculated by Hales' method are greater than those obtained in Figs. 6.13 and 6.14. An interpretation tech-nique based on a common point on the refractor allows a deeper penetration than one based on a common surface point. The arc (a) is obtained using the velocity for the compact rock in the calculations. For the lower arc (b) an average bedrock velocity of 3500 m/s has been used. Note again that the depths must be plotted in relation to the base of the top layer. The depth (b) here agrees rather well with depth (b) in Fig. 6.13.

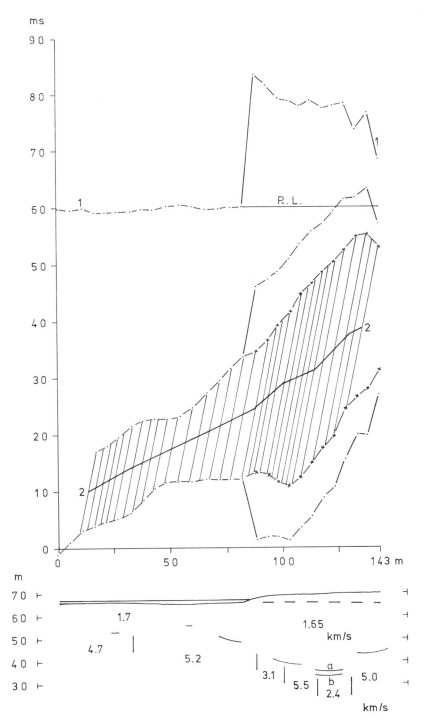

Fig. 6.15 Hales' method applied to data in Fig. 6.13.

Figure 6.16

The figure illustrates a four-layer case where there is a relatively small velocity contrast between the bottom refractor and the overlying layer. For the depth determination the recorded arrival times have been corrected by the ABC method (crosses) as well as by the ABEM method (dots). The ABC-curve used for the corrections is marked (3), at the top of the time–distance graph. This curve refers to the bottom refractor. The mean velocities in the latter layer are to be found in the mean-minus-T curve (1). Curve 2 giving the velocities in the third layer is obtained by a combination of corresponding velocity segments from shots B–E. The velocities given by curve 2 are those in a weathered and fractured rock layer.

The corrected travel times yield a clearer velocity picture than the raw travel times and now the slopes of the various velocity lines can be determined. However, there is also a tendency in the corrected travel times for the velocity in the third layer to increase with depth. There is probably no sharp boundary between the third and fourth layers but a gradual transition instead. At point D the velocity 1600 m/s is not detectable in the travel time curve but in the interpretation shown I have assumed that it exists.

Figures 6.17 and 6.18

At a cursory glance, the travel time curves in Fig. 6.17 appear to be parallel, except for the segments close to the shotpoints. A closer study reveals, however, that the curves from the distant shots and those from the shotpoints on the profile converge, but at a distance varying between 100 and 160 m from the shotpoints the curves are parallel to each other. The recorded travel times are plotted in Fig. 6.17 while the interpretation details are shown in Fig. 6.18.

The mean velocities in the bottom refractor are given by curve 5 in Fig. 6.18 while the corresponding recorded travel time curves are marked (1) and (2). Since curve 5 gives an average velocity of 5000 m/s, a correction line with this inverse slope has been used at all shotpoints. The correction procedure itself is not demonstrated but the result, namely the corrected travel times, is plotted in Fig. 6.18. It is now easier to discern the various velocity layers. Overlying the bottom refractor there is a layer where the velocity varies between 3300 m/s and 4600 m/s. The latter layer is more weathered and/or fractured than the underlying limestone. The upper layers display a more irregular velocity pattern. The velocities 2100–3100 m/s correspond to highly weathered and fractured limestone and the velocities 1200–1600 m/s to soil and disintegrated limestones.

In the graph in Fig. 6.18, curves 1 and 2 correspond to the bottom refractor while curves 3 and 4 consist of velocity segments from the overlying layer. The converging tendency between the curves from the two layers is now clearly visible. Curves 5 and 6 show the mean velocities in the bottom refractor and overlying layer respectively. The velocity distributions are shown in the cross-section.

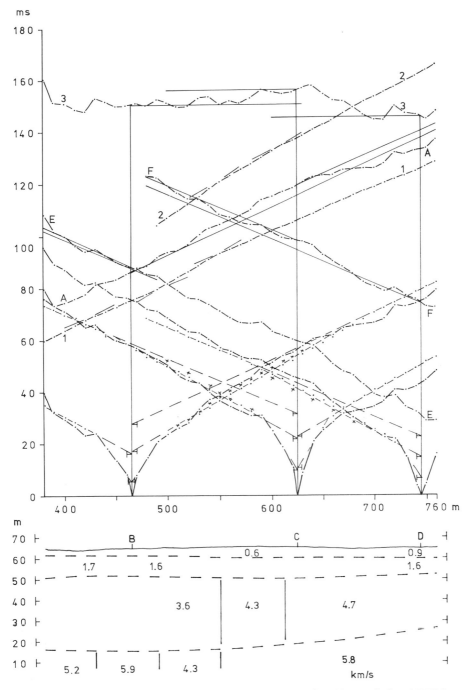

Fig. 6.16 Field example with small velocity contrast (Archives of the ABEM Company, Sweden).

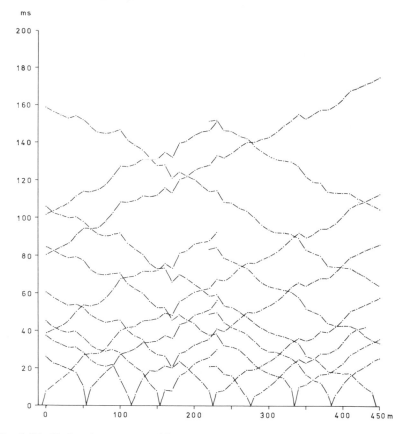

Fig. 6.17   Refraction curves with apparent parallelism (Archives of the ABEM Company, Sweden).

Schjöttelvik, Norway, Figures 6.19–6.23

This seismic investigation was carried out on behalf of the Norwegian State Power Board, NVE-Statskraftverkene, for a power scheme, machine station and tunnels. The investigation programme as a whole, comprising drilling, geotechnics and seismics, was co-ordinated by the Norwegian Geotechnical Institute, NGI.

The travel time curves in Fig. 6.19 indicate a very deep depression in the bedrock; the down-dip rock velocities are very low while the up-dip velocity segments are 'negative'. A conventional interpretation approach to the problem will grossly underestimate the depths. The profile length available is too short in relation to the depth of the bedrock and the rock is outcropping beyond either end of the profile. Various interpretation techniques are presented in Figs 6.19–6.22 and the results are summarized in the cross-sections in Fig. 6.23.

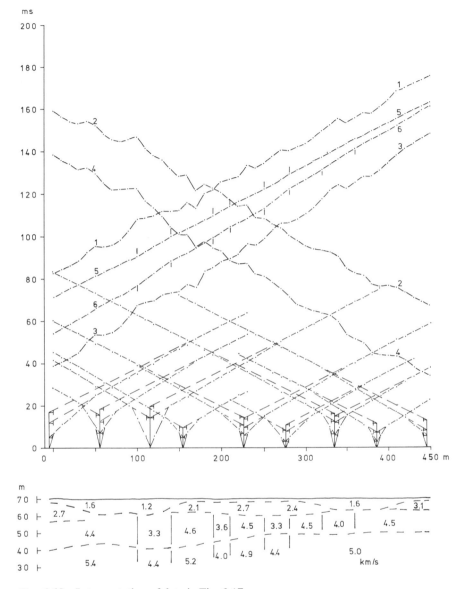

Fig. 6.18 Interpretation of data in Fig. 6.17.

A comparison in Fig. 6.19 between bedrock and overburden velocities shows that the velocity in the down-dip part of curve A is close to the velocity in the overburden layers. The curves M and N show at the end of the profile the overburden velocity in the area. The latter circumstance reveals that the slope of the right side of the depression is too great to allow registrations from

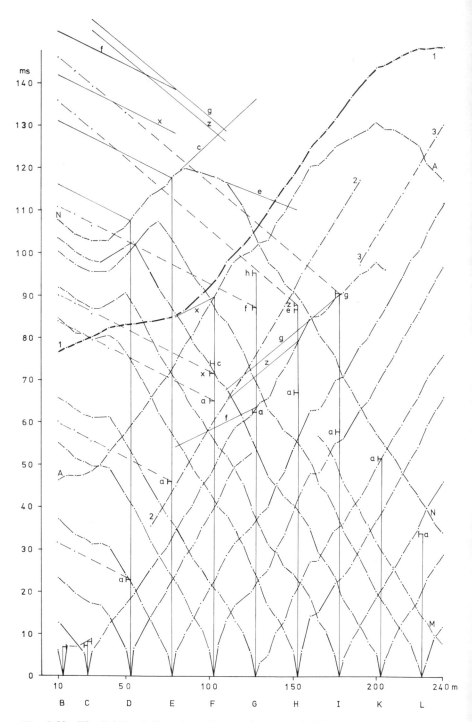

Fig. 6.19 The Schjöttelvik project, Norway (courtesy A/S Geoteam, Norway).

the bedrock. We also see that the velocity in the forward recording from C is somewhat lower than the velocity in curve A but higher than the overburden velocity. The higher velocity in curve A is probably due to a difference in the direction of the raypaths and the dip of the bedrock surface. The rays make an angle with the bedrock surface and therefore the waves from shot A are non-critically refracted into the overburden. On the other hand, the raypaths from point C are more nearly parallel to the surface of the bedrock.

The calculated velocities of the overburden layer, below a top layer of some few metres, is given by curves 2 and 3. This velocity is about 1500 m/s, except for a short section between points H and K where the velocity is 1600 m/s. The mean-minus-T method has been used in the usual way to obtain the velocities. The rock velocity obtained from curve 1 displays the characteristic features corresponding to a deep depression in the refractor, namely a central section with an extremely low velocity surrounded by too high velocities. The low-velocity segment in curve 1 probably reflects the combined influence of the depression and a shear zone in the bedrock.

The main problems are to obtain an estimate of the level of the deeper part of the bedrock and to locate any eventual shear zone in the bedrock. The information from the drill hole close to shotpoint L facilitates the depth interpretation (see Fig. 6.23).

At shotpoints B and C the bedrock lies at a depth of a few metres so that the waves are refracted back to the ground at a very short distance from the shotpoints. Note that the low velocity in the forward recording from point C is an apparent bedrock velocity and not an overburden velocity. The bedrock velocity segments in the reverse recording from shotpoints D, E and F have been adjusted by means of curve N and correction lines with the reciprocal slope of 5000 m/s. The velocity 5000 m/s has been chosen since the first part of curve 1 gives about 5200 m/s for the rock velocity. For the depth determinations at the points G, H, I, K, and L, the law of parallelism has been applied to curve N and to the respective velocity segments for the shotpoints. To adjust the overburden velocity segments, correction lines with the inverse slope of 1500 m/s and an overlying overburden curve have been used (the ABEM method). These lines are omitted in the figure for the sake of clarity. For the depth calculations themselves the velocities 1500 m/s and 5000 m/s have been used in the formulae. The velocities in the topmost layer, which is a few metres thick, vary between 400 m/s and 800 m/s.

The calculated depths are marked (a) in the upper part of Fig. 6.23. The letter (b) on the cross-section refers to the case when the ABC method has been used for the depth calculations. The marking (a/b) has been employed when the two methods have given identical results. The letters (c), (d), (e), etc. indicate modifications of the interpretation techniques. These are dealt with later on in connection with Figs 6.19–6.22.

It can be seen that the depth ascertained by drilling (Fig. 6.23) is considerably greater than that obtained by employing a conventional interpretation technique at the shotpoints, the arcs marked (a) in Fig. 6.23. However, methods described in Chapter 4 can be used in order to get better estimates of the real depths. From a study of the depth interpretation in Fig. 6.19 it appears that the depth calculations at points B–E are acceptable. The bedrock velocity segments in the reverse shooting from points D and E and the curve N, used for the corrections, refer to the same high-velocity rock section at the beginning of the profile. The problem starts at shotpoint F. In the reverse recording from F, the rays strike the bedrock within the section where the velocity is 5200 m/s, while the part of curve N that lies above F is probably affected by the lower velocity in the centre of the depression. The parallelism between the curves is lost because of this. On the graph in Fig. 6.19, the negative slope of curve N has been produced back to a point above shotpoint F. The law of parallelism has then been applied to the recorded curve N, the construction line and the bedrock velocity segment from F. The new intercept time and the corresponding depth are marked (c) in Figs 6.19 and 6.23 respectively. Note that the difference between the intercept times (a) and (c) corresponds to that between the recorded curve N and the construction line above point F. Another correction technique is to use a modification of the ABEM method. From the intersection of curve A and the vertical line through point F, a correction line (x) is drawn towards the beginning of the profile. The slope of the correction line is the reciprocal of 5000 m/s. From the point where this line meets the vertical line through the seismometer station immediately to the right of shotpoint E, a distance equal to the time difference between curves A and N at this point is laid off vertically upwards. From the point thus located, a line of the same slope (the reciprocal of 5000 m/s but proceeding upwards towards the left is drawn. This line is also denoted by (x) and each point on the bedrock velocity segment from F is lifted up by a distance equal to the distance between the correction line (x) and curve N. The intercept time (x) thus obtained coincides rather well with (c). The depth at point G is reinterpreted in the same way as for point F. Thus a correction line (f) is applied to the curve obtained from point C. At the station immediately to the right of E, a new correction line is drawn as before in relation to the curve from N. The depth obtained from this procedure is shown by the arc (f) on the upper cross-section in Fig. 6.23. The law of parallelism can also be used to estimate the depth at point G. An extension of the part of curve N with the negative slope, shown by line (c), back to point G yields a depth (h) about 7 m greater than found by using the intercept time (f).

The arc (d) below shotpoint H refers to a depth calculation where the same intercept time as for depth (a) has been used but the velocity 5000 m/s has been replaced in the depth formula by an assumed velocity of 3000 m/s, since the point H probably lies within a low-velocity zone in the bedrock. The calculated depth (d) is, very likely, also an underestimate because the rays from H strike the left flank of the depression, while the recorded arrival times in curve N, used for the corrections, refer to the right flank. The lower arc (e/z) below point H is based on two different modified interpretation techniques, namely the ABEM method and the law of parallelism. The correction line (z) is drawn backwards and at a point to the right of F a new line is constructed with reference to curve N in the manner described previously. Since shotpoint H lies above a supposed low-velocity zone, the slope of the correction line (z) corresponds to an assumed velocity of 3000 m/s in the rock. The intercept time and the depth designated (e) are obtained by a construction based on the law of parallelism. It seems likely that the bedrock velocity segments above E and F in curve N and above H and K in the forward curve A correspond to less steeply sloping bedrock surface and/or lower rock velocities. Therefore, a line (e) has been connected to the velocity segment on curve N vertically above points E and F and extended with the same slope to shotpoint H. The law of parallelism applied partly to the recorded curve N and the bedrock velocity segment from H and partly to the line (e) yields the intercept time (e). Since the intercepts (e) and (z) are almost of the same magnitude, a common depth has been calculated, marked (e/z) in Fig. 6.23. The depth (g) below shotpoint I has been obtained using the same assumptions and technique as for point H.

The reinterpreted depths in the deeper parts of the depression agree better with the depth obtained by drilling and with the general picture of the depths according to the ABC-curve in Fig. 6.20. The greater intercept times in the central part of the traverse indicate that the depths have to be greater than at the place of the drill hole where the intercepts are considerably less (Fig. 6.20). However, it should be noted that constructions and the use of curves from the opposite direction for corrections involve a number of assumptions and the uncertainty in the calculated depth increases with increasing distance from the shotpoint until at last the parallelism between the curves used for the interpretation ceases to exist. Therefore, from a theoretical viewpoint, less significance can be attached to the depth (g) at shotpoint I.

The depth determinations according to the ABC method are given in Fig. 6.20. Curves 1 and 2 refer to the bedrock while curve 3 corresponds to the sum of the times through the upper overburden layer. In order to facilitate the depth calculations, curve 3 is placed on the figure with the same reciprocal time, R.T., as curves 1 and 2. Curve 3 has been completed at the beginning of the profile by the calculated intercept time at shotpoint C. Because of the shallow depth involved, there are no registrations here from the layer with the velocity 1500 m/s. The ABC-curve (2) for the bedrock is based on the reciprocal time between shotpoints C and K and has been constructed by means of the curves from C, K and N, except over the intervals between 10 and 40 m and 200 and 240 m on the profile where arrival times have been read from curve A. Curve 1 is based on the arrival times recorded in curves A and N. For comparison this curve has been tied to curve 2. At the beginning of the profile curve 1 gives intercept times that are too large because of the previously mentioned fact that the waves from shotpoint A are probably non-

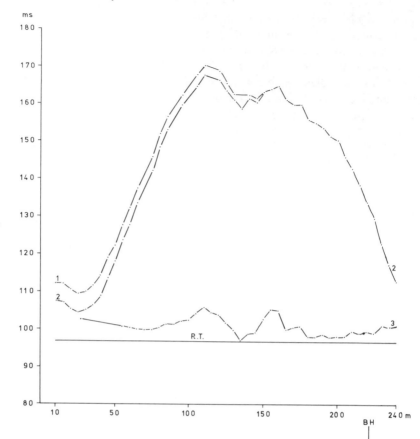

Fig. 6.20   Velocity analysis of data in Fig. 6.19 by the ABC method.

critically refracted into the overburden. As can be seen in Fig. 6.19, the curves from A and C approach each other.

The depths obtained by the ABC method differ but little from those determined at the shotpoints when a velocity variation in the bedrock is not considered. These depths are marked (a/b) on the upper cross-section in Fig. 6.23. A rather large discrepancy can be seen at shotpoint L where the depth marked (b) is that obtained by the ABC method. Here, there is a considerable difference between the depth obtained by drilling and the seismically determined depths because the bedrock surface is too steeply inclined to allow reliable depth determinations. If in the central part of the depression account is taken of a possible lower rock velocity, the depths obtained by the ABC method will be greater. The velocities 1500 m/s and 3000 m/s, applied to the intercept time at point H for the second overburden layer and bedrock respectively, give a depth coinciding with (d) in Fig. 6.23, as is to be expected since the intercept times at this point are identical in Figs 6.19 and 6.20.

The known depth, 46.9 m, in the drill hole and the ABC-curve in Fig. 6.20 can be used to estimate the depth in the centre of the traverse. The difference in time on the ABC-curve between the location for the drill hole and the arbitrarily chosen point 140 m along the profile gives an additional depth of about 25 m, i.e. a total depth of 72 m, as shown by the arc (i) in Fig. 6.23.

An application of Hales' method adds valuable information to the evaluation of the depths as well as the rock velocities. The interpretation using this method is to be found in Fig. 6.21. The time loop is based on the travel time curves from shotpoints C and K, duly completed by curves A and N. The ABC-curve (1) for the overburden has been used to correct the recorded arrival times of the two limbs of the time loop. In order to maintain the reciprocity for the arrival times from points C and K, the correction lines have been placed on the ABC-curve above K and C for the forward and reverse recordings respectively. The corrected travel times are shown by crosses. It should be noted, however, that the correction technique used does not eliminate the influence of the uppermost layer with the velocity 400–800 m/s. The velocity and thickness conditions in the top layer at the shotpoints between which the time loop is closed are postulated to be valid for the entire profile. In this particular case, since the upper layer is rather thin, an average weighted velocity will be 1470 m/s instead of 1500 m/s, the latter being the velocity in the second overburden layer. This difference in velocity is negligible.

The complete term $V_1 \sin i_{12} \cos \varphi_1 / \cos (\varphi_2 - \varphi_1)$ has been used for calculating the inclination of the slope lines. The dip angle φ_2 is estimated from the previously calculated depths. The width of the low-velocity zone in curve 2 is less than that in Fig. 6.19. Moreover, there is in Fig. 6.21 an indication that the shear zone is composed of two rock sections with different velocities. Between points F and G the velocity is 3100 m/s and in the larger part of the zone the computed velocity is about 2000 m/s. It should be noted, however, that the velocities in the zone as computed in Figs 6.19 and 6.21 are probably too low. They reflect not only the conditions in the shear zone but also a longer travel path in the bedrock as compared with the horizontal distance in the time plot.

The depths determined by Hales' method are given in the lower cross-section in Fig. 6.23. The arcs marked (a) are based on calculations in which the lower velocity in the centre of the depression is not taken into account, i.e. for the depth formula the velocities 1500 and 5000 m/s have been used to calculate the angle of incidence i_{12}. The depths (a) on the lower cross-section in Fig. 6.23 are about 10 m greater than the corresponding depths (a/b) in the upper cross-section where the same assumption concerning the bedrock velocity was made. The fact that Hales' method yields depths greater than those obtained by the intercept times at the shotpoints and geophone stations clearly indicates that the calculated depths are underestimated. A velocity of 3000 m/s in the shear zone has been assumed in calculating the depths denoted

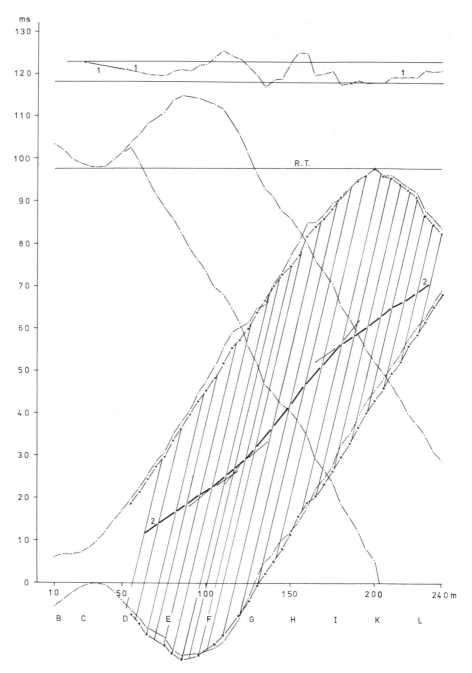

Fig. 6.21 Hales method applied to data in Fig. 6.19.

by (b). These depths give approximately the same bedrock surface as (f) and (e/z) in the upper cross-section. The slope lines used for the depth determinations are not shown in Fig. 6.21. They are based on the simplified term $V_1 \sin i_{12}$, i.e. the influence of the dips has not been taken account of, as recommended by Hales. The centres of the depth arcs have been plotted in relation to a line drawn between the points C and K.

In Fig. 6.22 the interpretation techniques described in Section 4.1.6 have been used to evaluate the depths in the centre of the depression. The travel time curves from shotpoints C and K are shown in the plot. These curves have been completed by arrival times from the offset shots A and N (see Fig. 6.19) in the usual manner. The depths can be determined in two ways. Before calculating the depths, the recorded arrival times can be corrected by means of curve 1, the ABC-curve for the overburden layer below the topmost layer, or the original arrival times can be used directly for the depth determinations. In the former case, the velocity 1500 m/s is to be used for the calculations and the depths obtained have to be plotted from the base of the top layer. The corrected arrival times are shown in the figure by crosses. When the recorded arrival times are employed without corrections, as is the case in Fig. 6.22, the average weighted velocity 1470 m/s should be used. The depths obtained are reckoned in this case from the ground surface.

The points 2, 3 and 4 in Fig. 6.22 are obtained when the interpretation technique of displacing the travel time curves is used. On the basis of earlier depth determinations let us assume a depth of 60 m and a rock velocity of 2500 m/s in the deeper part of the section. The term $2h_1 \tan i_{12}$ shows that one curve has to be displaced 90 m in relation to the other curve. In Fig. 6.22 the reverse arrival time at the coordinate 90 m on the profile has been moved 90 m to the right and added to the time at this point in the forward recording. The resulting time is plotted as point 2. In order to keep the figure to a reasonable size, the reciprocal time R.T. has been placed on the O-line. The intercept time given by point 2 when used in the equation $h_1 = T_2 V_1 \cos i_{12} / 2 \cos \varphi_2$ gives a value of about 73 m for h_1. The angle φ_2 was estimated from Fig. 6.23 using the assumed depth 60 m. The depth 73 m is then used to estimate new values for the dip angle φ_2 and the displacement. The time sum is now given by point 3. The recalculation of the depth yields about 80 m. A calculation based on an overestimate of the depth, 90 m, results in the same depth, namely 80 m. The corresponding intercept time is given by point 4. In this case the displacement is 135 m. If we assume that the rock velocity is approximately 3000 m/s instead of 2500 m/s, the calculated maximum depth will lie around 86 m. The depth values 80 and 86 m are shown in the upper cross-section in Fig. 6.23 by (k) and (l). The depths are plotted beneath the point 145 m along the profile, i.c. they have been moved half the displacement distance to the left.

The construction of lines 5 and 6 in Fig. 6.22 is based on the same reasoning as before for the intercepts at the shotpoints: see line (e) in Fig. 6.19. The sum of the constructed times is given by point 7 at a point 145 m along the profile.

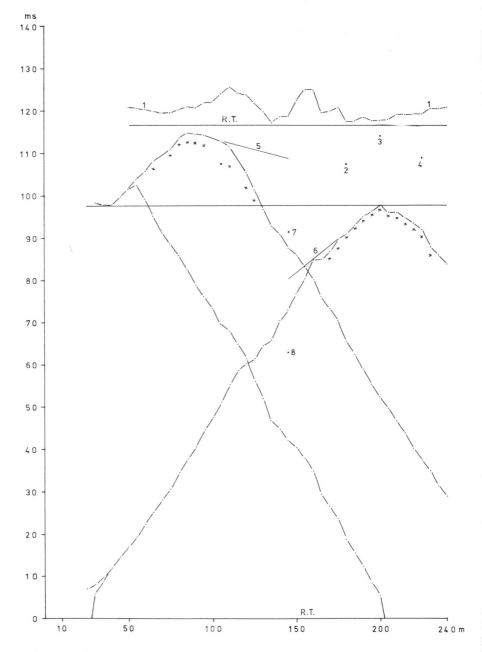

Fig. 6.22 Technique of Section 4.1.6 applied to data in Fig. 6.19.

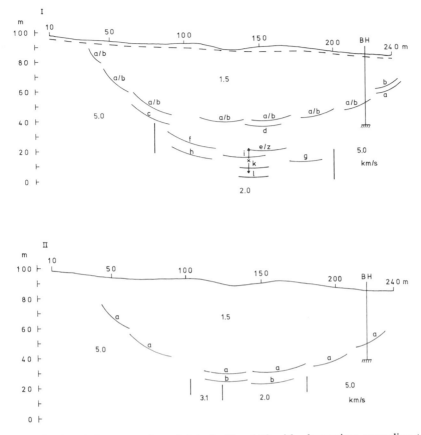

Fig. 6.23 Depth interpretation of data in Fig. 6.19 with alternatives according to various methods.

The equation to be used for the depth calculation is $h_1 = T_2V_1/2\cos(\varphi_2 - i_{12})$. The depth calculated in this manner is about 68 m and agrees with the depth marked (e/z) in Fig. 6.23.

A third way to evaluate the maximum depth is to sum the recorded arrival times and then employ the equation $h_1 = T_2V_1/2\cos(\varphi_2 + i_{12})$. The sum of the times is marked by point 8 in Fig. 6.22. The corresponding calculated depth is about 79 m, i.e. approximately the same as (k) in Fig. 6.23. In these calculations the sine function to obtain i_{12} is based on the velocities 1500 m/s and 5000 m/s.

A reasonable conclusion to be drawn from these rather lengthy calculations is that we can accept a depth of 75 m with an estimated tolerance of $\pm 10\%$ in the lower part of the bedrock relief. The estimated position of the bedrock surface is marked by a cross and the tolerance by arrows in the upper cross-section in Fig. 6.23.

The boundaries between the various rock sections obtained by the mean-minus-T method are shown in Fig. 6.23 I and those obtained by Hales' method in Fig. 6.23 II. The actual boundaries probably lie in between those indicated.

The interpretations could have been made easier by an extension of the profile to the right and by measuring one or two extra profiles perpendicularly

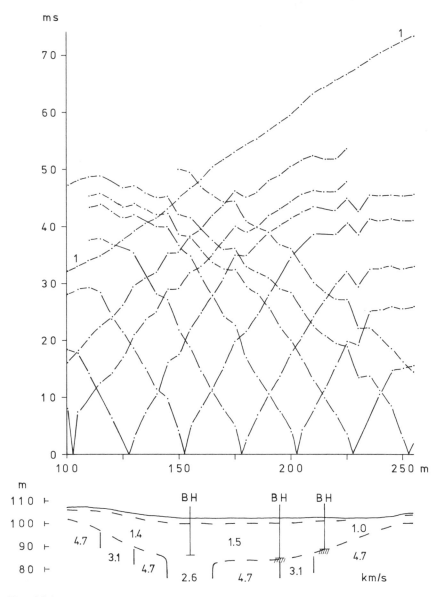

Fig. 6.24 Another profile in the Schjöttelvik project, Norway (cf. Fig. 6.19) (courtesy A/S Geoteam, Norway).

to the first profile. Nevertheless, even a single profile may give sufficient information for a feasibility evaluation of the project although the results are to some extent correspondingly uncertain.

Schjöttelvik, Norway, Figs 6.24 and 6.25

This seismic profile belongs to the project described earlier in connection with Figs 6.19–6.23. The geological picture is, however, considerably simpler here. The intercept times at the shotpoints were established by the ABEM method. The bedrock velocity analysis was carried out by the mean-minus-T method, curve 1 in the graph in Fig. 6.24. In the upper dry layer the velocities vary between 500 and 800 m/s. In the second overburden layer, mainly consisting of clay, the velocity is 1400–1500 m/s. The difference in velocity between the two overburden layers is due to a higher content of water in the second layer. The seismic interpretation was checked by drilling, except that the drill hole at a point 155 m on the profile was discontinued before the bedrock was reached. Hales' method has been used in Fig. 6.25 to determine the velocity distribution in the bedrock. Curve 1, which is the ABC-curve for

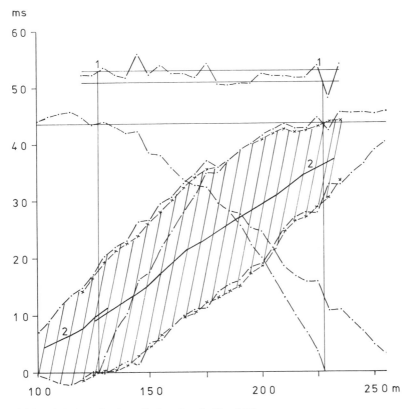

Fig. 6.25 Hales' method applied to data in Fig. 6.24.

the second overburden layer, has been used to correct the arrival times from the bedrock and the crosses indicate the corrected times. The velocity segments shown by the centre line, namely curve 2, yield more or less the same distribution as in Fig. 6.24. However, there are some discrepancies. The low-velocity zone at 125 m along the profile is less pronounced in Fig. 6.25 than in Fig. 6.24 and the zone around 200 m on the profile has a position in Fig. 6.25 to the right of the position in Fig. 6.24.

Nordingrå, Sweden, Figs 6.26 and 6.27

The object of this investigation was to locate a water well for the Municipality of Nordingrå. The rock of the area, gabbro, is known to have a rather low fracture and joint frequency so that locating a well with sufficient water yield using geological reasoning alone was not considered feasible.

The seismic survey showed two low-velocity zones. The probable directions of the zones are shown on the map, Fig. 6.26, by broken lines. The borehole, placed on the main low-velocity zone close to profile 3, gave more than 12 000 litres per hour (l/h). The time–distance graph, the cross-section and the data from the borehole are presented in Fig. 6.27. The bedrock was encountered in the borehole at a depth of 15.5 m. The fractured part of the bedrock is marked by a thick line along the borehole in Fig. 6.27. Compact rock was encountered at −37 m. The zone of crushed rock is probably dipping

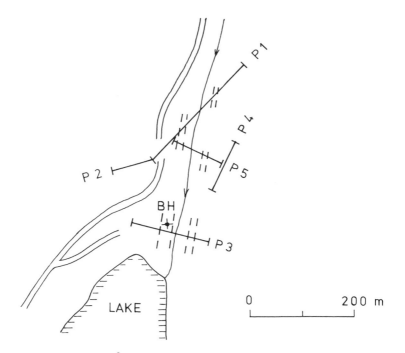

Fig. 6.26   The Nordingrå area, Sweden (Archives of the ABEM Company, Sweden).

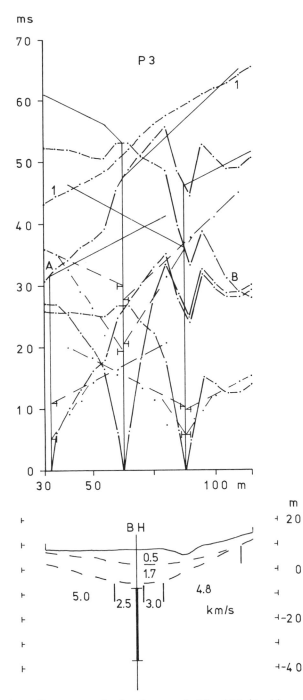

Fig. 6.27 Time–distance graphs for the area in Fig. 6.26 (Archives of the ABEM Company, Sweden).

and the term compact rock refers in this case to the footwall. The bedrock velocity distribution, shown by curve 1 in Fig. 6.27, is obtained by the mean-minus-T method applied to curves A and B.

Lidingö, Sweden, Fig. 6.28

The object of the investigation in the figure was to solve a hydrogeological problem. The drill holes A and B had yielded only 140–225 l/h at a drilled depth of about 120 m, while the holes C and D yielded 1100–1350 l/h at a depth of 60 m.

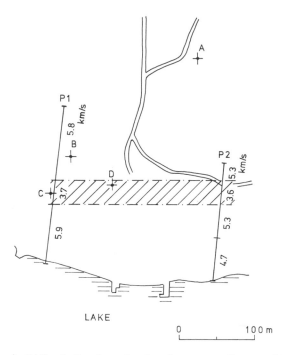

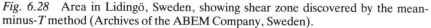

Fig. 6.28   Area in Lidingö, Sweden, showing shear zone discovered by the mean-minus-T method (Archives of the ABEM Company, Sweden).

The seismic survey was carried out to check if the technique of assessing rock quality by means of the mean-minus-T method, used since 1956, could explain the situation. The two profiles measured showed that the drill holes C and D happened to be located within a water-bearing, fractured rock section in which the seismic velocity was only 3600–3700 m/s as compared with 5800 m/s in the compact rock in which well B was drilled.

Märsta, Sweden, Figs 6.29 and 6.30

Another example of a seismic investigation for water prospecting is shown in Figs 6.29 and 6.30. The site is at Söderby, Märsta, west of Stockholm. The

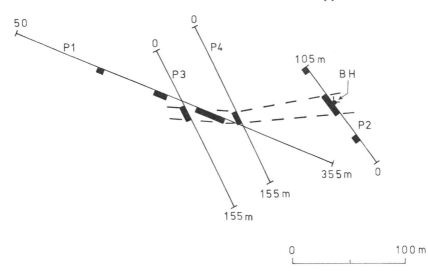

Fig. 6.29 Example of water prospecting by the seismic method (Märsta, Sweden) (Archives of the ABEM Company, Sweden).

bedrock consists of gneiss and is covered by a glacial till, except in a small part of the area of interest where it is outcropping. The flat terrain gives no clear indications of suitable locations for water wells, although the soil layer is rather thin. Therefore, a structural investigation by seismics was considered indispensable.

Three rock sections with relatively low velocities were detected along the first profile. The direction and extent of the main zone were checked by profiles 2–4 (Fig. 6.29). The water well was finally placed on profile 2. The cross-section and the drill hole are shown in Fig. 6.30. The main low-velocity zone is only 20 m wide and could not have been detected without a seismic

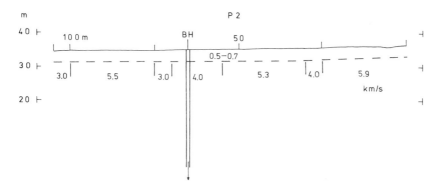

Fig. 6.30 Seismic cross-section along profile 2 in Fig. 6.29 (Archives of the ABEM Company, Sweden).

survey. The zone consists of two sections with different velocities, namely 3000 m/s and 4000 m/s. In estimating the possibilities of finding water in the bedrock, we do not only consider the low velocity as such but also the contrast between the velocity in the zone and the velocities in the surrounding more compact rocks. A convenient estimate of the velocity contrast is the ratio of the velocity in the solid rock to the velocity in the shear zone. In this particular case, the contrast is 1.4 and is considered promising. The number is the ratio of an average velocity 5600 m/s to the predominant velocity 4000 m/s in the zone.

An important problem when locating water wells in the bedrock is the determination of the dip of the shear zones. In the present case, the borehole was placed in the centre of the zone, since the zones in this region were assumed to be vertical. The well, drilled depth 120 m, gave 11 200 l/h after continuous test pumping for 17 days. The water level was then stabilized at a depth of 39 m. The natural groundwater level in this area is a few metres below the ground surface.

Since water wells placed at random in this area rarely yield more than 2000 l/h, the utility of a seismic survey is obvious. It has been found that the yield of seismically located water wells is 5–10 times that of an average well.

Skien River, Norway, Figs 6.31–6.33

The investigation described below in brief was carried out across the Skien River for a proposed fresh water tunnel under the river. It was suspected that somewhere in the river there was a prominent breccia zone separating the sedimentary rocks of the Oslo field on the one hand and the igneous rocks gneiss to the west on the other hand. It was feared that this contact zone between the two rock formations would create difficulties for the tunnel driving.

The critical section of the bedrock was initially detected along profile 1. Its position is marked on the map in Fig. 6.31 by a thick line. The strike of the low-velocity zone, shown by the dashed lines on the map, was determined by the additional profiles 2, 4 and 5. In order to find the location of the shear zone at the level of the proposed tunnel as well as the rock quality in it, an inclined borehole was placed on land with its azimuth along profile 5. The results from the drilling and the survey along profile 5 are shown in Fig. 6.32. Three low-velocity zones were observed along profile 5. The largest of these, velocity 2500 m/s, consists partly of consolidated breccia and partly of very loose alum shales. In the latter material the core losses were 75% on average. The two minor low-velocity zones proved to correspond to fractured sections. Water permeability tests carried out in the borehole gave values of 3–4 l/min in Lugeon indices, in rock sections where the velocity is 5500 m/s, and about 14 l/min in the main low-velocity zone. The permeability tests on the smaller fractured zones in the gneiss failed because of total water loss.

Eventually the tunnel was driven in the direction of profile 1. The tunnel

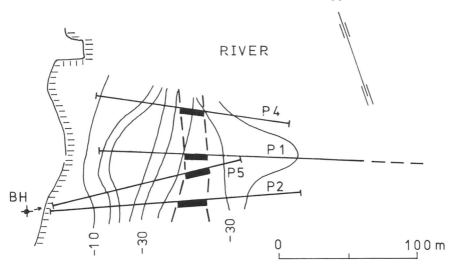

Fig. 6.31 Site map of the Skien River project, Norway and results of seismic investigation (Archives of the ABEM Company, Sweden).

and the results from profile 1 are shown in Fig. 6.33. The weaker rock section in the gneiss indicated by the velocity 3000 m/s gave some problems so that grouting was necessary. The rock conditions were further investigated by a horizontal borehole from the tunnel level before driving the tunnel through the main zone. The tunnel was reinforced over a length of about 12 m where the rock consists of a fractured zone and the alum shales mentioned above. A proper combination of drilling and seismics made it possible in this case to overcome the construction difficulties caused by very poor rock quality and a water head of 60 m.

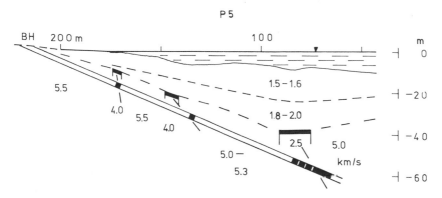

Fig. 6.32 Seismic cross-section and borehole along profile 5 for the area in Fig. 6.31 (Archives of the ABEM Company, Sweden).

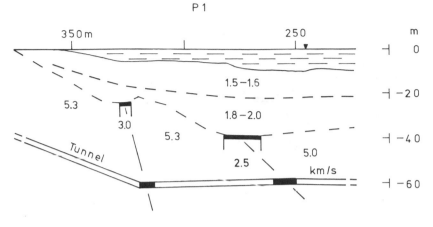

Fig. 6.33 Cross-section along profile 1 in Fig. 6.31 showing seismic results and tunnel (Archives of the ABEM Company, Sweden).

Letzi, Sweden, Figs 6.34 and 6.35

One of the first deliberate attempts to combine information from detailed rock velocity determinations by means of the mean-minus-T method and core drilling was made in 1958 for a hydroelectric power scheme at Letzi in northern Sweden. The drilling operations and seismic survey were commissioned by the Swedish State Power Board.

Drilling was started to check the rock quality in a prominent low-velocity zone detected in profile 205 in the preliminary seismic interpretation. The additional profiles 207 and 208 were measured to find an alternative site for the proposed tunnel and showed the zone to be continuous although less pronounced on the flanks of profile 205.

The seismic results along profile 205 and the drilling data are projected in Fig. 6.35 on to the section through the line A–A in Fig. 6.34. The boundaries of the shear zone, dipping at an angle of about 45° to the south, delineate the rock section where the fracturing frequency is greater than 20 cracks per metre. The log of the holes indicates that there is a gradual decrease in the fracturing intensity from the zone to the surrounding more solid granite, a not uncommon feature in such cases. Ultimately, it was decided to select an alternative tunnel site outside this region to avoid the difficulties due to intensely fractured rock sections.

Sundsbarm, Norway, Fig. 6.36

The low-velocity zones in Fig. 6.36 correspond in part to contact zones separating different rock types. Difficulties were caused during the tunnel driving by water flows in these zones. The rock velocity distribution was found by the mean-minus-T method.

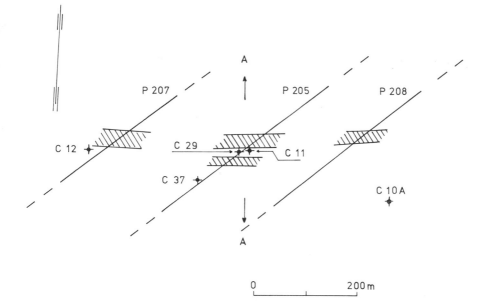

Fig. 6.34 Fracture zones detected by the mean-minus-T method in the Letzi area, Sweden (Archives of the ABEM Company, Sweden).

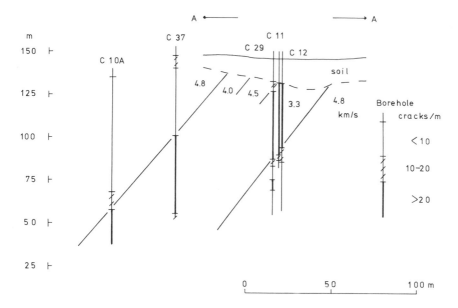

Fig. 6.35 Cross-section along line A–A in Fig. 6.34 (Archives of the ABEM Company, Sweden).

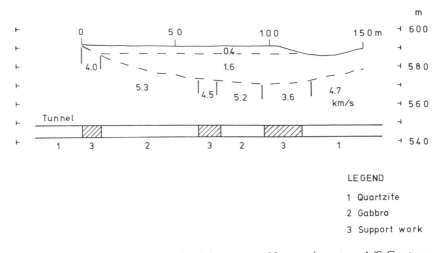

Fig. 6.36   Seismic results in the Sundsbarm area, Norway (courtesy A/S Geoteam, Norway).

Even though the velocity 4500 m/s is not remarkably low in relation to the velocities of the surrounding compact rocks, support work proved to be necessary. In general, when the velocity in the solid rock is 5000 m/s or above, a velocity equal to or lower than 4000 m/s indicates a section that can be critical for excavation work. However, it is even necessary sometimes to reinforce tunnel walls when the velocity contrast between the higher and lower velocities is less than the above. Since there is a gradual change in the rock quality, a sharp velocity limit cannot, however, be established. An intermediate velocity may be present in a section with alternating rock quality and the weaker parts determine the need of support work. Other factors of importance in this connection are, for instance, the depth of the excavation work below the bedrock surface and the strike of the zones in relation to the tunnel direction.

The Oslo field, Figs 6.37 and 6.38

In Fig. 6.37 a case is presented where the interpretations underestimated the depths in a local depression in the bedrock. The bedrock surface based on the original depth interpretation (made by the author in 1972) is shown by a dashed line in the cross-section. To the right of shotpoint C the calculated depths deviate considerably from those obtained in the drill holes, the maximum error being 28% below the profile coordinate 105 m.

The seismic profile is a part of a larger measuring programme that was carried out on behalf of the Norwegian Geotechnical Institute, NGI, in connection with a waste and sewage water project in the Oslo area for Vestfjord Avlöpsselskap, VEAS.

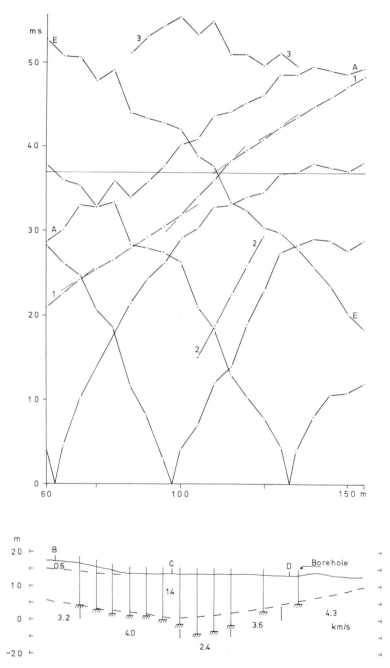

Fig. 6.37 Example of a seismic interpretation with initially underestimated depths and reinterpretation (Oslo field, Norway) (courtesy A/S Geoteam, Norway).

The soil consists of water-saturated clay except for an upper layer of sand and gravel, the thickness of which varies between 2.0 and 2.5 m. The drilling proved that the bedrock is directly overlain by clay so that the discrepancy mentioned above cannot be ascribed to a hidden layer with a higher velocity, for instance, a hard moraine layer. The velocity, 1400 m/s, in curve 2 is probably the velocity in the second overburden layer down to the bedrock. This velocity was determined by means of the mean-minus-T method applied to the arrival times from points C and D.

The bedrock velocity determined by means of the mean-minus-T method, curve 1 in Fig. 6.37, gives two velocity segments in the centre of the traverse, 2400 m/s and 3600 m/s at the distances 100–115 and 115–130 m respectively. Hales' method, Fig. 6.38, indicates one velocity segment, 2900 m/s, between 105 and 125 m along the profile (curve 1). The higher velocities in the area vary from 4000 to 4400 m/s. It seems likely that the extremely low velocity 2400 m/s in curve 1 in Fig. 6.37 is due to a combined effect of the local depression in the bedrock configuration and the shear zone.

The ABC-curve (3) conveys the impression that the greatest depth is to be found at 100 m along the profile and on both sides the intercept times decrease. In order to get a better estimate of the depths in the depression, I have utilized some of the interpretation methods presented in Section 4.1.6, namely the modified ABC method and the displacement method.

For the modified ABC method we have to use the equation $h_1 = T_2 V_1 / 2\cos(\varphi_2 + i_{12})$. The intercept time T_2 is found to be about 0.0178 s at 105–110 m in the profile. Other pertinent data are $V_1 = 1400$ m/s, $V_2 = 3000$ m/s and $i_{12} = 27.8°$. A value of about 8° for the average dip angle φ_2 is obtained from the first depth calculations. The first step in the successive calculations gives the depth 15.4 m. Using this depth we get about 12° for the dip angle φ_2. A renewed calculation leads to the depth 16.2 and a corresponding value of about 14° for φ_2. In the third calculation the depth has increased to 16.7 m. As mentioned in Section 4.1.6(a)(i), an overestimate of the depth leads more rapidly to a stable value for the depth. An assumed depth of 20 m results in a calculated depth of 18.5 m. In the next step the depth will be 17.9 m, which is almost the same depth as that obtained by the drilling at 105 m in the profile.

The technique of curve displacements can also be utilized. If curve E is connected to the bedrock velocity segment from shotpoint D and then moved 20 m to the right, the sum of the arrival times in curves C and D gives an intercept time 0.024 s. The corresponding depth calculated from the relevant equation $h_1 = T_2 V_1 \cos i_{12} / 2\cos\varphi_2$ will be about 15.0 m. If the calculation of the angle of incidence i_{12} is based on 1400 m/s and 3500 m/s the calculated depth is about 16.0 m. The average velocity 3500 m/s may be used in the calculations since the arrival times in the direct recording emanate from the rock section to the left of the zone where the velocity is 4000 m/s. An average of the depths 17.9, 15.0 and 16.0 m gives an estimated depth within 0.5–1.0 m of the true one in the depression.

Assessment of the rock quality and recommendations for the support works needed in the tunnel were made by NGI. The bedrock consists of limestone and shale belonging to the sedimentary formations of the Oslo field. The lower rock velocities in the seismic profile correspond to clay filled joints and an eruptive dyke with fissured contact zones with the surrounding sedimentary rocks. The support work in the tunnel in the actual region ranges from shotcrete, steel rib shotcrete to bolted steel rib reinforced shotcrete.

The estimated tunnel stability was also expressed by NGI in numbers as the rock mass quality Q. The value of Q is a function of six parameters according to the equation

$$Q = \frac{RQD}{J_n} \frac{J_r}{J_a} \frac{J_w}{SRF}$$

where

RQD = rock quality designation (Deere, Peck, Monsees and Schmidt, 1963)

J_n = joint set number

J_r = joint roughness number

J_a = joint alteration number

J_w = joint water reduction factor

and

SRF = stress reduction factor

The factor RQD/J_n gives a measure of the relative block size while the factors J_r/J_a and J_w/SRF refer to the shear strength and active stress respectively. For further information a paper by Barton, Lien and Lunde (1974) should be consulted.

According to these authors the numerical value of Q ranges from 0.001 (for exceptionally poor quality squeezing-ground) up to 1000 (for exceptionally good quality rock which is practically unjointed).

In the actual case the value of Q varies between 1 and 5 and the corresponding velocity is about 3000 m/s (Fig. 6.38). The rock quality is classified as poor to fair. In the region where the velocity is 4400 m/s, the Q value was estimated to be 20 and support work was not needed.

The Oslo field, Figs 6.39 and 6.40

Profile 11 shown in Fig. 6.39 is from the same project as the example of Fig. 6.37. The seismically determined bedrock configuration agrees in this case with the depths subsequently obtained by drilling. A perfect match is not to be expected since the drill holes lie 15 m to the side of the seismic line. The low-velocity zone in profile 11 was also detected in profile 15 (Fig. 6.40). The

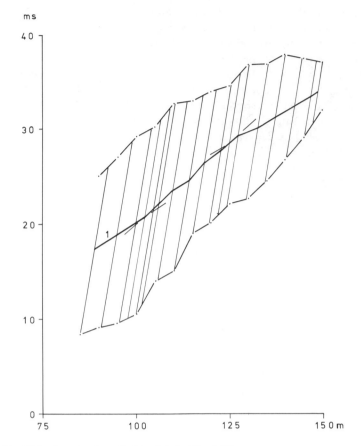

Fig. 6.38 Hales' method applied to data in Fig. 6.37.

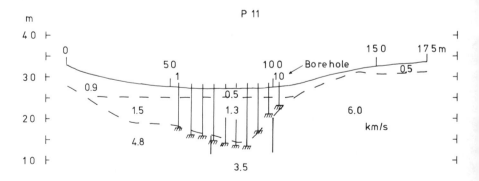

Fig. 6.39 Another seismic section in the project in Fig. 6.37 (courtesy A/S Geoteam, Norway).

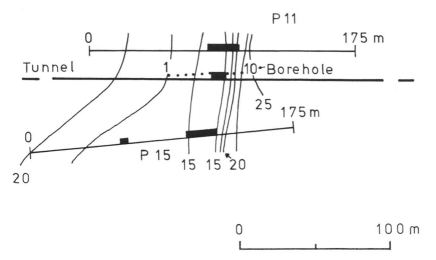

Fig. 6.40 Bedrock contours for the area in Fig. 6.39 (courtesy A/S Geoteam, Norway).

zone coincides with a depression in the bedrock. The bedrock contour lines 20, 15, etc. in Fig. 6.40 are based on the seismic and drilling results. The drill holes were placed in the tunnel line where the seismic investigations had indicated a depression in the bedrock surface. The geological examination of the tunnel, carried out by NGI, showed that the low-velocity zone corresponds to a weathered zone containing clay and gravel. The material in the zone was classified as earthy. The calculated Q value is very low, namely 0.012. The tunnel support had to be in the form of a cast concrete arch. The low velocities and the support work in the tunnel are shown in Fig. 6.40 by thick lines.

Tiruchirapalli, India, Figs 6.41 and 6.42

Figures 6.41 and 6.42 show one of the profiles measured in co-operation with the Tamil Nadu Water Supply and Drainage Board (TWAD Board) during a training course in refraction seismics.

The corrected arrival times indicate that there are four layers overlying the compact rock. The velocity pattern was repeated in three other profiles from the area. The interpretation was also confirmed by a nearby water well. The correlation between velocity layers and geological formations encountered in the well is as follows:

500 m/s	Soil
1700 m/s	Highly weathered biotite gneiss
2800 m/s	Weathered biotite gneiss
3500 m/s	Jointed biotite/granite gneiss
4900–5400 m/s	Biotite gneiss

The velocity determination in the bottom refractor (fifth layer) is shown in curve 1 in Fig. 6.41. The corresponding velocity analysis when Hales' method is used is presented in Fig. 6.42. In the latter interpretation the velocity lines (curve 2) are more irregular owing to the fact that the recorded, uncorrected arrival times have been used. Neither of the curves indicates the presence of any outstanding rock section of poor quality.

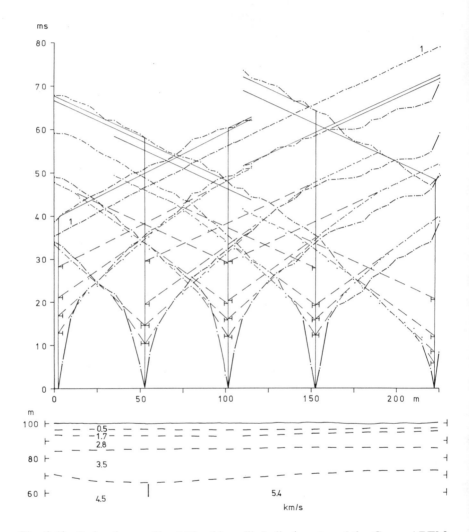

Fig. 6.41 Refraction profile at Tiruchirapalli, India (courtesy Atlas Copco ABEM AB, Sweden).

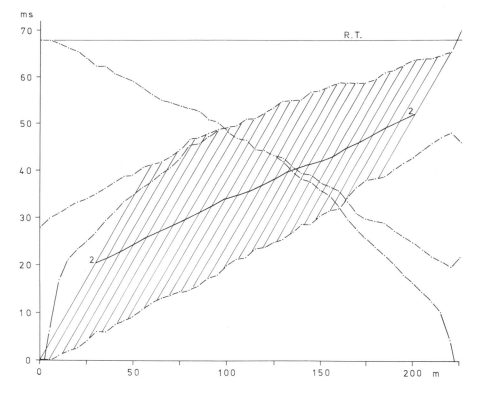

Fig. 6.42 Hales' method applied to data in Fig. 6.41.

El Marj, Libya, Figs 6.43–6.46

Large-scale geophysical surveys comprising seismic and earth resistivity investigations have been carried out in connection with the agricultural programme undertaken by the Executive Authority of Jabel El Akhdar in El Marj area, the Socialist People's Libyan Arab Jamahiria. Sections of one of the seismic profiles are shown in Figs 6.43–6.46 and discussed below.

The average shotpoint distance in these measurements was 220 m and a geophone spacing of 20 m was considered to be suitable. It was thus possible to cover a line of length 460 m with 24-channel equipment and at the same time to obtain a sufficient number of arrival time recordings from the various velocity layers. The continuous method was employed with the arrival times from the various geophone spreads tied to each other by means of overlapping geophone stations. The offset shots had to be extended 400–600 m from the geophone spreads. In order to increase the energy delivered to the ground, the shots were fired at depths varying from 3 to 5 m in 4-in drill holes. Every second shotpoint is omitted in the graphs.

The travel time curves of Fig. 6.43 show the velocity pattern within the area

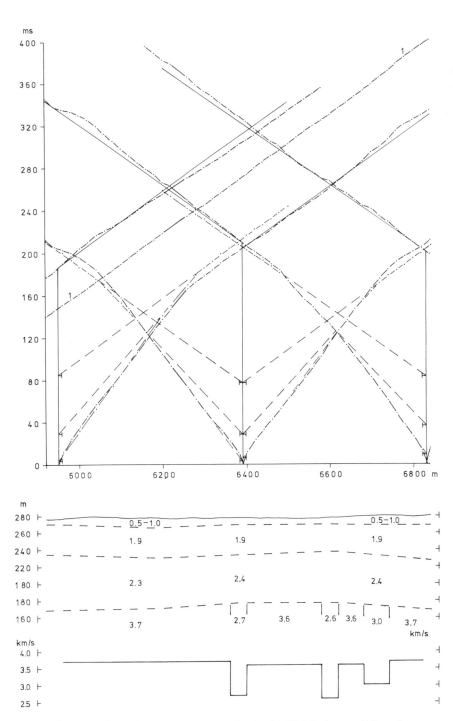

Fig. 6.43 Part of a refraction seismic profile in the El Marj area, Libya (courtesy BEHACO, Sweden).

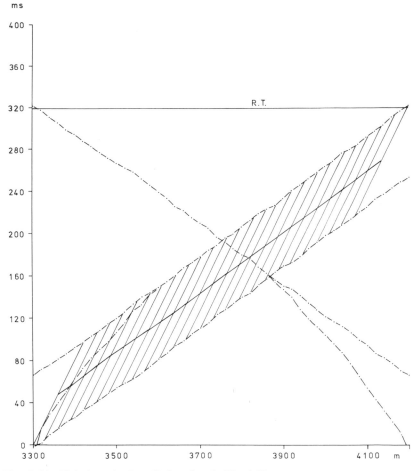

Fig. 6.44 Hales' method applied to data in Fig. 6.43.

very clearly. The various velocity segments present an almost ideal case and corrections to the recorded arrival times are hardly needed. The upper layer, velocity 500–1000 m/s, corresponds to the soil type 'terra rossa'. The next layer in which the recorded velocity is about 1800 m/s has the same composition as the top layer but contains more water. The velocities 2300 m/s and 2400 m/s are obtained in Quaternary limestone breccia, redeposited sandstone, sandy marl and terra rossa. The bottom layer, velocities 3000–3700 m/s, consists of Eocene sediments, classified as dolomitic, very marly limestone.

Because of the considerable depths to the Eocene limestone and the magnitude of the angles of incidence, the waves strike the ground 60–70 m from the point of emergence on the limestone surface. Therefore, the mean-minus-T method provides only a general velocity picture and the details are missing. This can be seen in curve 1 in Fig. 6.43. Even if curve 1 cannot give the velocity distribution in the bottom layer, it is needed as an intermediate step for the evaluation of the correction lines and for the depth calculation. For a detailed velocity analysis Hales' method has to be applied to the problem (Fig. 6.44). Since we have to deal with a multilayer case, it has to be transformed into a simple two-layer case. A fictitious, weighted velocity V_x for the overburden layers was obtained by using the equation $\Sigma_1^{n-1} V_v d_v = V_x \Sigma_1^{n-1} d_v$ according to Section 4.2.2. The term $V_x \sin i_{xn}$ gives the slope lines to be used. The centre line through the midpoints on the slope lines in Fig. 6.44 indicates that there are rather small velocity variations in the limestone. The velocities found are plotted in the cross-section in Fig. 6.43.

It is difficult to discern the various velocity line segments in the recorded curves in Fig. 6.45. The corrected times give a better picture of the velocities. In Fig. 6.46 the velocities in the Eocene limestone, the fourth layer, have been determined by Hales' method. In this case the recorded arrival times have been corrected by means of curve 3, which is the ABC-curve for the second overburden layer. The crosses denote the corrected times. The centre line (2) indicates three sections with lower velocities and the velocity analysis in curve 4 yields a similar velocity pattern. The latter curve is obtained by displacing the reverse travel time curves from the bottom layer 140 m to the right and using the mean-minus-T method for the velocity calculation. Curve 1 has then been moved 70 m to the left to get the position of the velocity segments. The velocities obtained from curve 2 in Fig. 6.46 are shown in the cross-section in Fig. 6.45. Beneath the section the velocities are also plotted on the vertical axis as a function of distance. Curve 1 in Fig. 6.45 gives the mean velocity in the Eocene limestone when the mean-minus-T method has been applied to the undisplaced curves.

Stiegler's Gorge, Tanzania, Figs 6.47 and 6.48

Figures 6.47 and 6.48 show the interpretation of one of the seismic profiles measured for a proposed dam site in connection with the Stiegler's Gorge Power Project on the Rufiji River, Tanzania. The investigations were made on behalf of Rufiji Development Authority, Rubada, and the Norwegian Agency for International Development, NORAD, under the supervision of A/S Hafslund, Norway.

The bedrock consists of alternating, nearly horizontal strata of Permian and Triassic sandstones and mudstones. The upper parts of the bedrock are partly severely weathered, causing a diffuse transition between soil and rock layers. The soil, ranging from clayey, silty sand to sandy, clayey silt, is in general a few metres thick.

Curves A and B in Fig. 6.47, which are records of the arrival times of waves from the bottom layer, are parallel at the beginning of the profile, but between 70 and 90 m along the profile the parallelism is lost. The time

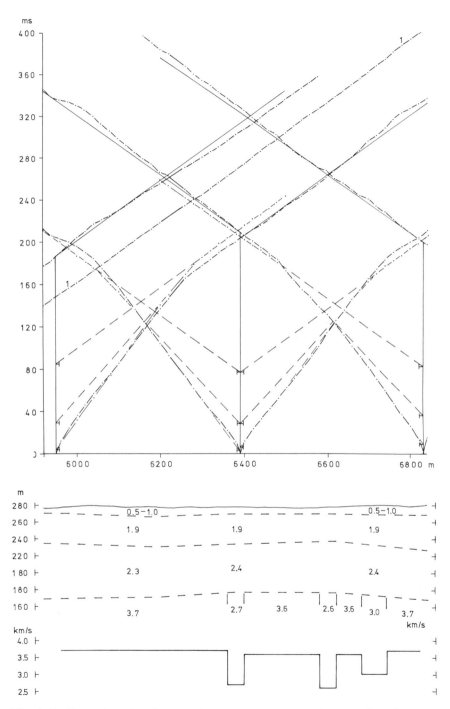

Fig. 6.45 Part of a refraction seismic profile in the El Marj area, Libya (courtesy BEHACO, Sweden).

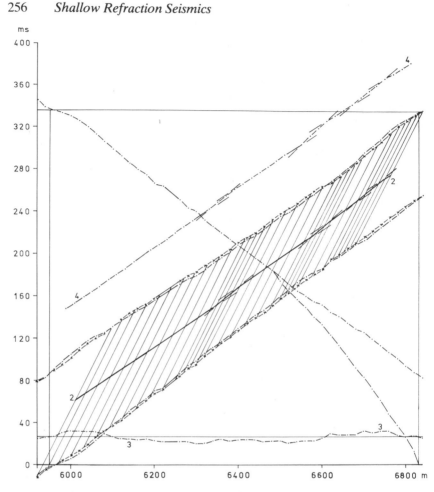

Fig. 6.46 Hales' method applied to data in Fig. 6.45.

increments between adjacent geophones are greater in curve B than in curve A. It seems likely that the waves from the more distant shot A have penetrated a deeper part of the bedrock where the velocity is locally higher. The curve obtained by the mean-minus-T method using the arrival times in curve A and the reverse registrations is marked 1 and indicates a low-velocity segment between 50 and 90 m. When the curve from shotpoint B, instead of that from A, is employed for the velocity determination the calculated velocity in the zone is found to be lower, curve 2.

The ABEM method was used to correct the recorded arrival times for the depth determinations, with correction lines based on the reciprocal slope of 3000 m/s.

The velocity analysis for the bottom layer according to Hales' method is

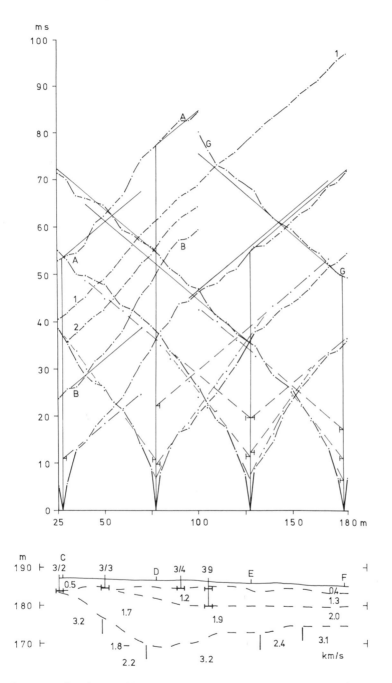

Fig. 6.47 A refraction profile in the Stieglers' Gorge Project, Tanzania (courtesy A/S Geoteam, Norway).

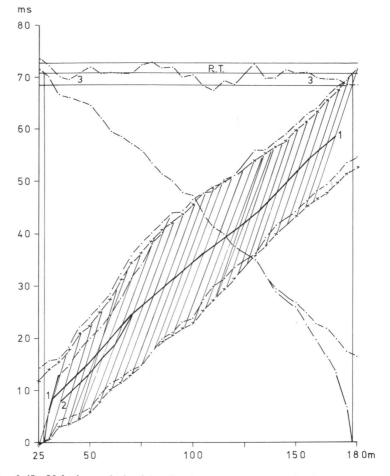

Fig. 6.48   Velocity analysis of data in Fig. 6.47 by Hales' method.

shown in Fig. 6.48. The raw arrival times have been adjusted by means of curve 3, which is the ABC-curve for the strata below the top layer. The crosses denote the corrected times. The slope lines have to be altered along the traverse because of the varying depths and velocities. At the beginning of the profile they correspond to 750 m/s, in the centre to 950 m/s and from 125 metres onwards to 850 m/s. The presumed low-velocity zone is clearly seen in curve 1. There is also in curve 1 an indication, approximately between 130 and 150 m along the profile, of a segment with a lower velocity. This lower velocity is not detectable in the velocity determination in Fig. 6.47 where curve 1 yields a general velocity picture in the bottom layer. Because of the small velocity contrast between the layers, the points where the waves strike the ground surface are displaced 10 to 15 m in relation to the points where they leave the surface of the bottom layer. Curve 2 in Fig. 6.48 is for an interpretation in which the arrival times from point C have been used. As can be seen, the velocity is lower in the upper part of the zone. The velocity distribution in the bottom layer found from curve 1 in Fig. 6.48 is shown in the cross-section in Fig. 6.47.

Correlation between velocity layers and the composition of the material of the upper layers could be made in test pits 3/2, 3/3, 3/4 and 39. The soil, velocities 400–600 m/s, consists of sandy, silty clay. In all test pits the ripping stopped on sandstone, marked by short lines on the cross-section. The sandstone layer with the velocity 1200–1300 m/s is probably highly weathered and loose. This layer could be penetrated by auger boring performed at test pit 39. The drilling was stopped at the surface of the layer with the velocity 1900 m/s.

A certain relation was established between rock quality, expressed in RQD values obtained from bore cores, and seismic velocities. The correlation is given as curve 3 in Fig. 2.5 in Chapter 2. Judging from this analysis, a velocity in the bottom bedrock equal to or higher than 3000 m/s may correspond to moderately fractured rock with RQD value about 80%. The velocities in the zone, 1800–2200 m/s, indicate RQD values between 40 and 60% and a rock quality in the range fair to poor.

7

Future prospects

It is always difficult to predict future technical development but with the seismic refraction technique there are so many clues that I am rather convinced it can be done without risking appreciable discrepancies between reality and prediction. Modern digital recording equipment, the mechanical energy sources available, an increased interest in including elastic constants, frequencies and amplitudes in evaluations, and a tendency to replace manual treatment of seismic data by computer processing clearly indicate the development to be expected.

It is very likely that:

1. Explosives will be replaced by mechanical energy sources to a large extent. When explosives are used it is rather difficult to balance the energy introduced into the earth and the signal amplification so that we obtain the data desired, except for the first arrivals of the longitudinal waves. On the other hand, the combination of a digital recorder with stacking and filter possibilities and a mechanical energy source enables the geophysicist to break the recording when the desired information is obtained. Besides the first arrival times of the longitudinal waves, such information may be records of transverse waves, second longitudinal arrivals and reflections from the subsurface layers.

   However, the very existence of mechanical energy sources may reduce the applicability of the refraction method if the seismic profiles have to be placed in the terrain according to the adaptability of the energy source and not according to the real needs of the project. It may also be too tempting to propose a mechanical energy source in order to avoid all the problems connected with the use of explosives, and the survey may be a failure because of too weak signals.

2. The interpretation work will largely be made by computers, including the determination of the arrival times as well as the corrections and calculations of depths and velocities. The interpretation methods proposed in Chapter 4 can conveniently be computerized and, in fact, they partly are,

even today. As mentioned previously, the first arrivals have almost exclusively been used in refraction surveys. A contributory cause of this is that – in my opinion – the interpreters have simply had no time to carry out a complete analysis of the measuring data available, but, with the aid of modern seismic equipment and computer processing, more information will be extracted from the records with considerably less time and effort, and, moreover, various interpretation techniques can be applied to solve intricate problems.

However, computer programs can be used uncritically and an interpreter may rely too much on mathematics and lose contact with the geological realities. If the interpretations are made by computers it is indispensable that the plotted travel times are displayed on a viewing screen or on a printout for evaluation of, for instance, the accuracy in time plotting, the consistency between the various travel time curves, the parallelism between the curves and the number of velocity layers. The different steps in the interpretation work must also be visualized.

The applicability of the refraction method will probably increase still more in the future provided that modern equipment and computer processing are used, not to simplify the interpretations, but to increase the accuracy and reliability of the seismic results.

References

Atlas Copco ABEM AB (1983), ABEM TERRALOC seismic system, Bromma, Sweden.

Barry, K. M. (1967), Delay time and its application to refraction profile interpretation, in *Seismic Refraction Prospecting* (ed. A. W. Musgrave), SEG, Tulsa, pp. 348–61.

Barthelmes, A. J. (1946), Application of continuous profiling to refraction shooting, *Geophysics* **11**, 24–42.

Barton, N., Lien, R. and Lunde, J. (1974), Engineering classification of rock masses for the design of tunnel support, *Rock Mech.*, **6**, 180–286.

Baumgarte, J. (1955), Konstruktive Darstellung von seismischen Horizonten unter Berucksichtigung der Strahlenbrechung im Raum. *Geophys. Prosp.*, **3**, 126–62.

Deere, D. U., Peck, R. B., Monsees, J. E. and Schmidt, B. (1969), *Design of Tunnel Liners and Support Systems*, Department of Civil Engineering, University of Illinois, Urbana, USA.

Dobrin, M. B. (1976), *Introduction to Geophysical Prospecting*, McGraw Hill, New York.

Domzalski, W. (1956), Some problems of shallow refraction investigations. *Geophys. Prosp.*, **4**, 140–66.

Edge, A. B., and Laby, T. H. (1931), *The Principles and Practice of Geophysical Prospecting*, Cambridge University Press, London.

Gardner, L. W. (1939), An areal plan of mapping subsurface structure by refraction shooting. *Geophysics*, **4**, 247–59.

Gardner, L. W. (1949), Seismograph determination of salt-dome boundary using well detector deep on dome flank. *Geophysics*, **14**, 29–38.

Gardner, L. W. (1967), Refraction seismograph profile interpretation, in *Seismic Refraction Prospecting* (ed. A. W. Musgrave), SEG, Tulsa, pp. 338–47.

Hagedoorn, J. G. (1959), The plus-minus method of interpreting seismic refraction sections. *Geophys. Prosp.*, **7**, 158–82.

Hales, F. W. (1958), An accurate graphical method for interpreting seismic refraction lines. *Geophys. Prosp.*, **6**, 285–94.

Hawkins, L. V. (1961), The reciprocal method of routine shallow seismic refractions. *Geophysics*, **6**, 806–19.

Heiland, C. A. (1940), *Geophysical Exploration*, Prentice Hall, New York.

Jakosky, J. J. (1950), *Exploration Geophysics*, 2nd Edn, Trija, Los Angeles, Calif.

Layat, C. (1967), Modified Gardner delay time and constant distance correlation interpretation, in *Seismic Refraction Prospecting* (ed. A. W. Musgrave), SEG, Tulsa, pp. 171–93.

Meissner, R. (1965), *Wellenfrontverfahren für die Refraktionsseismik*, Ber. Univ. – Inst. Meteor. Geophys., Frankfurt 9.

Pakiser, L. C. and Black, R. A. (1957), Exploring for ancient channels with the refraction seismograph. *Geophysics*, **22**, 32–47.

Palmer, D. (1980), The generalized reciprocal method of seismic refraction interpretation (ed. Kenneth B. S. Burke), SEG, Tulsa.

Parasnis, D. S. (1984), Chapter 3 in *Water in Crystalline Rocks*, Unesco Publications, Paris, In press.

Rockwell, D. W. (1967), A general wavefront method, in *Seismic Refraction Prospecting* (ed. A. W. Musgrave), SEG, Tulsa, pp. 363–415.

Schenck, F. L. (1967), Refraction solutions and wavefront targeting, in *Seismic Refraction Prospecting* (ed. A. W. Musgrave), SEG, Tulsa, pp. 416–25.

Sheriff, R. E. (1973), *Encyclopedic Dictionary of Exploration Geophysics*, SEG, Tulsa.

Sjögren, B., Öfsthus, A. and Sandberg, J. (1979), Seismic classification of rock mass qualities. *Geophys. Prosp.*, **27**, 409–42.

Sjögren, B. (1979), Refractor velocity determination – cause and nature of some errors. *Geophys. Prosp.*, **27**, 507–38.

Sjögren, B. (1980), The law of parallelism in refraction shooting. *Geophys. Prosp.*, **28**, 716–43.

Tarrant, L. H. (1956), A rapid method of determining the form of a seismic refractor from line profile results. *Geophys. Prosp.*, **4**, 131–39.

Telford, W. M., Geldart, L. P., Sheriff, R. E. and Keys, D. A. (1976), *Applied Geophysics*, Cambridge University Press, London.

Thornburgh, H. R. (1930), Wavefront diagrams in seismic interpretation. *Bull. Am. Assoc. Petrol. Geol.*, **14**, 185–200.

Wyrobek, S. M. (1956), Application of delay and intercept times in the interpretation of multilayer refraction time distance curves. *Geophys. Prosp.*, **4**, 112–30.

Further reading

Alterman, Z., Censor, D., Ginzburg, A. and Schoenberg, M. (1974), The poor man's seismic source: a computer game. *Geophys. Prosp.*, **22**, 261–71.

Attewell, P. B. and Ramana, Y. V. (1966), Wave attenuation and internal friction as functions of frequency in rocks. *Geophysics*, **31**, 1049–56.

Avedik, F. and Renard, V. (1973), Seismic refraction on continental shelves with detectors on sea floor. *Geophys. Prosp.*, **21**, 220–28.

Bamford, D. and Nunn, K. R. (1979), In situ seismic measurements of crack anisotropy in the carboniferous limestone of northwest England. *Geophys. Prosp.*, **27**, 322–38.

Bandu Rao Naik, S., Nanda Kumar, G. and Vijaya Raghava, M. S. (1980), The correlation refraction method as applied to weathered zone studies in a granite terrain. *Geophys. Prosp.*, **28**, 18–29.

Beyer, M. G. (1968), Ground water production from the bedrock of Sweden, in *Ground Water Problems*, Proc. Int. Symp., Stockholm, 1966, Pergamon Press, Oxford, pp. 161–79.

Birch, F. (1966), Compressibility, elastic constants. *Handbook of Physical Constants* (ed. S. P. Clark), Geological Society of America, Memoir 97, pp. 97–173.

Bogoslovsky, V. A. and Ogilvy, A. A. (1970), Application of geophysical methods for studying the technical status of earth dams. *Geophys. Prosp.*, **18**, 758–73.

Brede, E. C., Johnston, R. C., Sullivan, L. B. and Viger, H. L. (1970), A pneumatic seismic energy source for shallow water/marsh areas. *Geophys. Prosp.*, **18**, 581–99.

Brooke, J. P. (1973), Geophysical investigations of a landslide near San Jose, California. *Geoexploration*, **11**, 61–73.

Bruckshaw, J. M. and Mahanta, P. C. (1961), The variation of the elastic constants of rocks with frequency, in *Early Papers*, European Association of Exploration Geophysicists, The Hague, 65–76.

Burke, K. B. S. (1973), Seismic techniques in exploration of quaternary deposits. *Geoexploration*, **11**, 207–31.

Butler, D. K. and Curro, J. R. (1981), Crosshole seismic testing – procedures and pitfalls. *Geophysics* **46**, 23–29.

Caterpillar Tractor Co. (1972), *Handbook of Ripping – a Guide to Greater Profit*.

Cecil, O. S. (1971), Correlation of seismic refraction velocities and rock support requirements in Swedish tunnels, Reprints and preliminary reports, No. 40, Swedish Geotechnical Institute, Stockholm.

Cole, D. I. (1976), Velocity/porosity relationships in limestones from the Portland Group of southern England. *Geoexploration*, **14**, 37–50.

Crampin, S., McGonigle, R. and Bamford, D. (1980), Estimating crack parameters from observations of P-wave velocity anisotropy. *Geophysics*, **45**, 345–60.

Cummings, D. (1979), Determination of depths to an irregular interface in shallow seismic refraction surveys using a pocket calculator. *Geophysics*, **44**, 1987–98.

Dampney, C. N. G. and Whiteley, R. J. (1980), Velocity determination and error analysis for the seismic refraction method. *Geophys. Prosp.*, **28**, 1–17.

Datta, S. (1967), Elastic measurements in rock formations. *Geoexploration*, **5**, 115–26.

Davis, A. M. (1978), *A Technique for the in situ Measurement of Shear Wave Velocity*, Atlas Copco ABEM AB, Bromma, Sweden.

Dix, C. H. (1955), Seismic velocities from surface measurements. *Geophysics*, **20**, 68–86.

Ewing, W. M., Jardetzky, W. S. and Press, F. (1957), *Elastic Waves in Layered Media*, McGraw-Hill, New York.

Faust, L. Y. (1951), Seismic velocity as a function of depth and geological time. *Geophysics*, **16**, 192–206.

Goguel, F. M. (1951), Seismic refraction with variable velocity. *Geophysics*, **16**, 81–101.

Green, R. (1962), The hidden layer problem. *Geophys. Prosp.*, **10**, 166–70.

Green, R. (1974), The seismic refraction method – a review. *Geoexploration*, **12**, 259–84.

Greenhalgh, S. A. and King, D. W. (1981), Curved raypath interpretation of seismic refraction data. *Geophys. Prosp.*, **29**, 853–82.

Gureev, A. M. (1967), Deformability of rock foundations of dams, in The National Committee on Large Dams, UDC 627.8:624.131.25.004, translated from *Gidrotekh. Stroit.*, No. 2, 52–6, February 1967.

Hamdi Falah, A. I. and Taylor Smith, D. (1981), Soil consolidation behavior assessed by seismic velocity measurements. *Geophys. Prosp.*, **29**, 715–29.

Hasselström, B., Rahm, L. and Scherman, K. A. (1964), Methods for the determination of the physical and mechanical properties of rock. Transactions of 8th Congress on Large Dams, Edinburgh, **1**, pp. 611–26.

Hasselström, B. (1969), Water prospecting and rock investigation by the seismic refraction method. *Geoexploration*, **7**, 113–32.

Hatherly, P. J. (1982), A computer method for determining seismic first arrival times. *Geophysics*, **47**, 1431–6.

Hawkins, L. V. and Maggs, D. (1961), Nomograms for determining maximum errors and limiting conditions in seismic refraction survey with a blind-zone problem. *Geophys. Prosp.*, **9**, 526–32.

Helbig, K. and Mesdag, C. S. (1982), The potential of shear-wave observations. *Geophys. Prosp.*, **30**, 413–31.

Helfrich, H. K., Hasselström, B. and Sjögren, B. (1970), Complex geoscience investigation programmes for siting and control of tunnel projects, in *The Technology and Potential of Tunneling*, Volume 1 (ed. N. G. W. Cook), Cygnet Print Ltd, Johannesburg.

Hobson, G. D. and Hunter, J. A. (1969), In-situ determination of elastic constants in overburden using a hammer seismograph. *Geoexploration*, **7**, 107–11.

Hunter, J. A. and Hobson, G. D. (1977), Reflections on shallow seismic refraction records. *Geoexploration*, **15**, 183–93.

Jankowsky, W. (1970), Empirical investigation of some factors affecting elastic wave velocities in carbonate rocks. *Geophys. Prosp.*, **18**, 103–18.

Johnston, D. H., Toksöz, M. N. and Timur, A. (1979), Attenuation of seismic waves in dry and saturated rocks: II Mechanisms. *Geophysics*, **44**, 691–711.

Kaila, K. L. and Narain, H. (1970), Interpretation of seismic refraction data and the solution of the hidden layer problem. *Geophysics*, **35**, 613–23.

Kaila, K. L., Tewari, H. C. and Krishna, V. G. (1981), An indirect seismic method for determining the thickness of a low-velocity layer underlying a high-velocity layer. *Geophysics*, **46**, 1003–8.

King, M. S. (1966), Wave velocities in rocks as a function of changes in overburden pressure and pore fluid saturants. *Geophysics*, **31**, 50–73.

Kitsunezaki, G. (1980), A new method for shear-wave logging. *Geophysics*, **45**, 1489–506.

Knox, W. A. (1967), Multilayer near-surface refraction computations, in *Seismic Refraction Prospecting* (ed. A. W. Musgrave), SEG, Tulsa, 197–216.

Laski, J. D. (1973), Computation of the time–distance curve for a dipping refractor and velocity increasing with depth in the overburden. *Geophys. Prosp.*, **21**, 366–78.

Leet, L. D. (1950), *Earth Waves*, Harvard University Press, New York.

Levin, F. K. (1980), Seismic velocities in transversely isotropic media, II. *Geophysics*, **45**, 3–17.

Linsser, H. (1961), The generation of seismic waves by explosions, in *Early Papers*, European Association of Exploration Geophysicists, The Hague, 54–64.

Lund, C.-E. (1974), The hidden layer in seismic prospecting, Geophysical Memorandum 7/74, Atlas Copco ABEM AB, Bromma, Sweden.

MacPhail, M. R. (1967), The midpoint method of interpreting a refraction survey, in *Seismic Refraction Prospecting* (ed. A. W. Musgrave), SEG, Tulsa, 260–6.

Meidav, T. (1967), Shear wave velocity determination in shallow seismic studies. *Geophysics*, **32**, 1041–6.

Meissner, R. (1961), Wave-front diagrams from uphole shooting. *Geophys. Prosp.*, **9**, 533–43.

Meissner, R. (1965), P- and SV-waves from uphole shooting. *Geophys. Prosp.*, **13**, 433–59.

Merrick, N. P., Odins, J. A. and Greenhalgh, S. A. (1978), A blind zone solution to the problem of hidden layers within a sequence of horizontal or dipping refractors. *Geophys. Prosp.*, **26**, 703–21.

Meyer, R. (1978), The continuous seismic refraction method. *Bull. Assoc. Eng Geol.*, **15**, 37–49.

Mooney, H. M. (1973), *Handbook of Engineering Geophysics*, Bison Instruments, Minneapolis, Minnesota, USA.

Morgan, N. A. (1969), Physical properties of marine sediments as related to seismic velocities. *Geophysics*, **34**, 529–45.

Mota, L. (1954), Determination of dips and depths of geological layers by the seismic refraction method. *Geophysics*, **19**, 242–54.

Musgrave, A. W. (ed.) (1967), *Seismic Refraction Prospecting*, SEG, Tulsa.

Nanda Kumar, G. and Vijaya Raghava, M. S. (1981), On the significance of amplitude studies in shallow refraction seismics. *Geophys. Prosp.*, **29**, 350–62.

Nelson, P. H., Magnusson, K. A. and Rachiele, R. (1982), Application of borehole geophysics at an experimental waste storage site. *Geophys. Prosp.*, **30**, 910–34.

Newman, P. J. and Worthington, M. H. (1982), In-situ investigation of seismic body wave attenuation in heterogeneous media. *Geophys. Prosp.*, **30**, 377–400.

Northwood, E. J. (1967), Nomogram for curved-ray problem in overburden, in *Seismic Refraction Prospecting* (ed. A. W. Musgrave), SEG, Tulsa, 296–303.

O'Brien, P. N. S. (1960), The use of amplitudes in refraction shooting – a case history. *Geophys. Prosp.*, **8**, 417–28.

O'Brien, P. N. S. (1961), A discussion on the nature and magnitude of elastic absorption in seismic prospecting. *Geophys. Prosp.*, **9**, 261–75.

Officer, C. B. (1958), *Introduction to the Theory of Sound Transmission*, McGraw-Hill, New York.

Olhovich, V. A. (1964), The causes of noise in seismic reflection and refraction work. *Geophysics*, **29**, 1015–30.

Overmeeren van, R. A. (1981), A combination of electrical resistivity, seismic refraction, and gravity measurements for groundwater exploration in Sudan. *Geophysics*, **46**, 1304–13.

Palm, H. (1981), An interpretation of seismic refraction data by use of travel-times, amplitudes and frequencies, Societas Upsaliensis pro Geologia Quaternaria, Uppsala, Sweden.

Palm, H. (1981), On the use of trace amplitudes and frequencies in seismic refraction investigations, Societas Upsaliensis pro Geologia Quaternaria, Uppsala, Sweden.

Parasnis, D. S. (1972), *Principles of Applied Geophysics*, Chapman and Hall, London.

Peet, W. E. (1960), A shock wave theory for the generation of the seismic signal around a spherical shot hole. *Geophys. Prosp.*, **8**, 509–33.

Peraldi, R. and Clement, A. (1972), Digital processing of refraction data – study of first arrivals. *Geophys. Prosp.*, **20**, 529–48.

Perkov, Yu. R. and Dolgikh, M. A. (1965), Experience in comparative determinations of the elasticity moduli of rocks in the laboratory and in the field, UDC 624.131.43 + 620.172.22, translated from *Osn., Fundam. Mekh. Gruntov*, No. 3, pp. 10–11, May–June 1965.

Poley, J. Ph. and Nooteboom, J. J. (1966), Seismic refraction and screening by thin high-velocity layers: a scale-model study. *Geophys. Prosp.*, **14**, 184–203.

Redpath, B. B. (1973), Technical report E-73-4, Seismic refraction exploration for engineering site investigations, US Army Engineer Waterways Experiment Station, Livermore, Calif.

Ricker, N. (1953), The form and laws of propagation of seismic wavelets. *Geophysics*, **18**, 10–40.

Saxena, P. C. and Ravendra Nath (1978), Application of geophysical methods to some typical engineering geology problems. *Proc. 3rd Int. Congr. International Association Engineering Geologists*, Section 4, Volume 2, pp. 56–64.

Schmöller, R. (1982), Some aspects of handling velocity inversion and hidden layer problems in seismic refraction work. *Geophys. Prosp.*, **30**, 735–51.

Scott, J. H. (1973), Seismic refraction modeling by computer. *Geophysics*, **38**, 271–84.

Sjögren, B. and Wager, O. (1969), On a soil and ground water investigation with the shallow refraction method at Mo i Rana, Norway. *Eng. Geol. (Amsterdam)*, **3**, 61–70.

Slotnik, M. M. (1950), A graphical method for the interpretation of refraction profile data. *Geophysics*, **15**, 163–80.

Soske, J. L. (1959), The blind zone problem in engineering geophysics. *Geophysics*, **24**, 359–65.

Stacy, T. R. (1976), Seismic assessment of rock masses. *Symposium on Exploration for Rock Engineering – Johannesburg*. Steffen, Robertson and Kirsten, Geotechnical Engineers, Johannesburg.

Stephanson, O., Lande, G. and Bodare, A. (1979), A seismic study of shallow jointed rocks. Pergamon Press, Oxford. *Rock Mech. Min. Sci. Geomech. Abstr.*, **16**, 319–27.

Stierman, D. J., Healy, J. H. and Kovach, R. L. (1979), Pressure induced velocity gradient: an alternative to a Pg refractor in the Gabilan range, central California. *Bull. Seism. Soc. Am.*, **69**, 397–415.

Stierman, D. J., and Kovach, R. L. (1979), An in situ velocity study: the Stone Canyon well (Calif.). *J. Geophys. Res.*, **84**, 672–8.

Stulken, E. J. (1967), Constructions, graphs and nomographs for refraction computations, in *Seismic Refraction Prospecting* (ed. A. W. Musgrave), SEG, Tulsa, 304–29.

Surendra Singh (1978), An iterative method for detailed depth determination from refraction data for an uneven interface. *Geophys. Prosp.*, **26**, 303–11.

Ukhov, S. B. and Panenkov, A. S. (1968), Relation between the static and the dynamic deformation indexes of rock in large-scale tests on rock masses, UDC 552.1:624.131.379, translated from *Gidrotekh. Stroit.*, No. 11, 33–7, November 1968.

Vijaya Raghava, M. S., Jawahar, G. and Sherbakova, T. V. (1977), An ultrasonic profiling investigation on some fresh and weathered granites of Hyderabad, India. *Geophys. Prosp.*, **25**, 768–79.

Vijaya Raghava, M. S. and Nanda Kumar, G. (1979), The blind-zone problem in multiple refraction-layer overburden situations. *Geophys. Prosp.*, **27**, 474–9.

Ward, R. W. and Hewitt, M. R. (1977), Mono-frequency borehole traveltime survey. *Geophysics*, **42**, 1137–45.

Wardell, J. (1970), A comparison of land seismic sources. *Geoexploration*, **8**, 205–29.

Whitcomb, J. H. (1966), Shear-wave detection in near-surface seismic refraction studies. *Geophysics*, **31**, 981–83.

Whiteley, R. J. and Greenhalgh, S. A. (1979), Velocity inversion and the shallow seismic refraction method. *Geoexploration*, **17**, 125–41.

Woolley, W. C., Musgrave, A. W. and Gray, H. (1967), A method of in-line refraction profiling, in *Seismic Refraction Prospecting* (ed. A. W. Musgrave), SEG, Tulsa, pp. 267–89.

Wright, C. and Johnson, P. (1982), On the generation of P and S wave energy in crystalline rocks. *Geophys. Prosp.*, **30**, 58–70.

Index

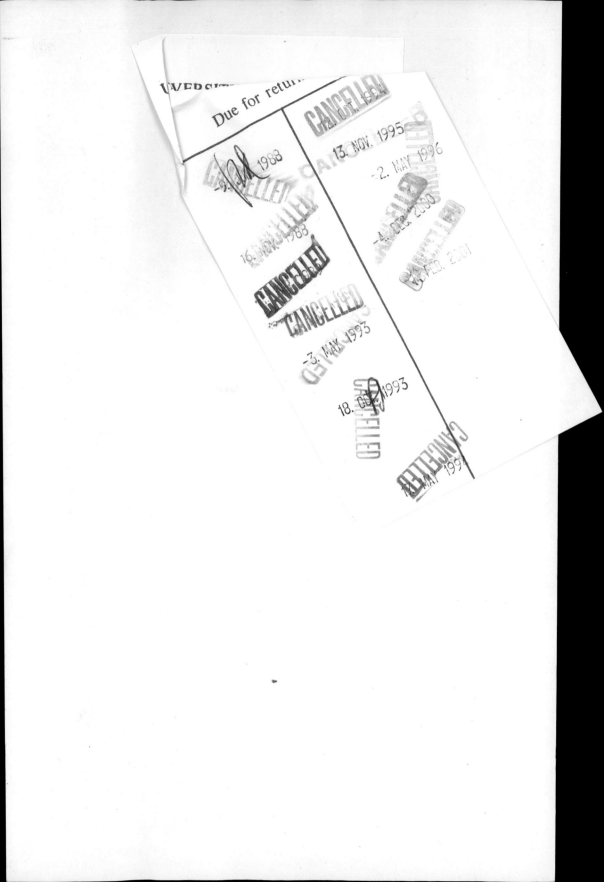